NELSON MATHS

AUSTRALIAN CURRICULUM **NSW**

Student Book

K

Glenda Bradley

NELSON
A Cengage Company

Australia • Brazil • Mexico • Singapore • United Kingdom • United States

Contents

Writing Numbers to 5

✡ Trace over the numbers.

1	1	1	1
2	2	2	2
3	3	3	3
4	4	4	4
5	5	5	5

✡ Trace over the numbers.

Bingo 5

✦ Play a game.

You will need:

- a dice with a ★ sticker covering the 6
- a partner

Make your bingo grid:

- Roll the dice and say the number to your partner.

- Your partner writes the number in a box below.

- Take turns rolling the dice and saying the number until every box has a number in it.

- If you roll a ★, you can choose 1 2 3 4 or 5.

How to play:

- In turn, roll the dice.

- Say the number. If you have that number in a box, colour it in.

- When you have coloured 5 numbers in a row, you win. Good luck!

Unit 1 **Numbers to 5** (TRB pp. 22–25)
Whole numbers MAe-4NA counts to 30, and orders, reads and represents numbers in the range 0 to 20

5

Dot-to-Dot

Join the numbers in the correct order.

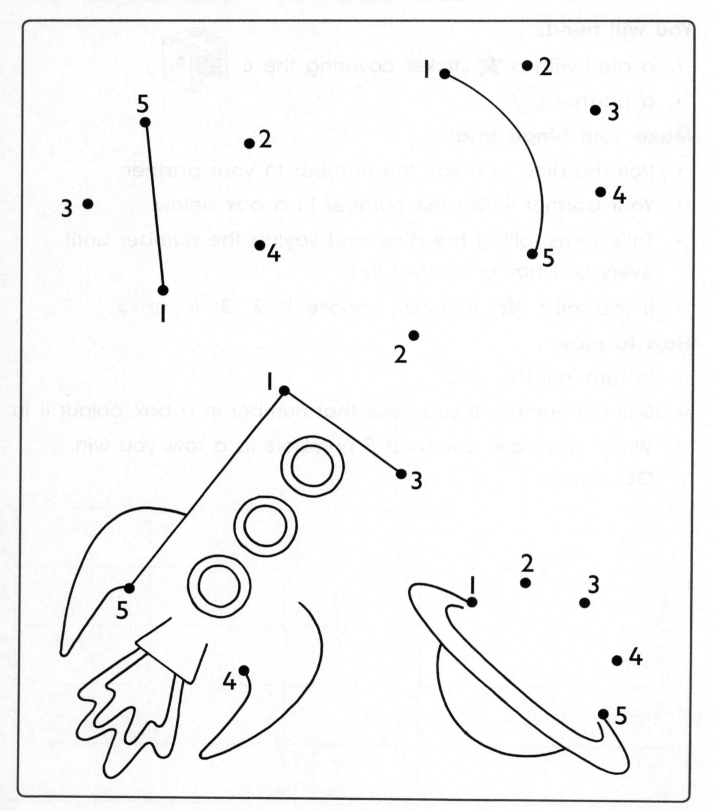

Unit
1 **Numbers to 5** (TRB pp. 22–25)
Whole numbers MAe-4NA counts to 30, and orders, reads and represents numbers in the range 0 to 20

STUDENT ASSESSMENT

✺ Count up to 5.

✺ Write the numbers in the correct order.

3 5 2 4 1

Write these numbers in the correct order, too.

5 3 1 4 2

✺ Start at 1 and join the dots.

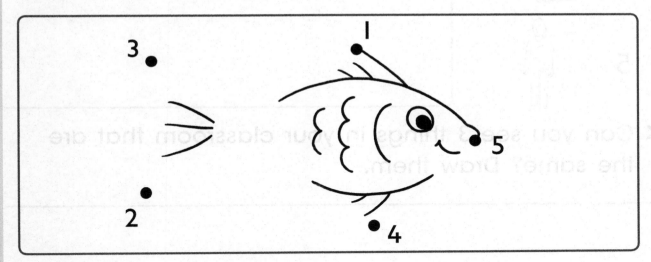

✺ Ask your partner to tell you a number.

Write it in the box.

✺ Have your partner check your number.

Unit
1
Numbers to 5 (TRB pp. 22–25)
Whole numbers MAe-4NA counts to 30, and orders, reads and represents numbers in the range 0 to 20

7

In the Classroom

✹ In the big box, draw:

2

4

1

3

5

✹ Can you see 3 things in your classroom that are the same? Draw them.

8

School Things

DATE:

✿ Draw the correct number of things in the box.

3

1

4

2

5

✿ Draw a circle around the number that matches the group.

1   2   3   4   5

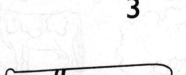

1   2   3   4   5

Unit 2   **Counting to 5** (TRB pp. 26–29)
Whole numbers MAe-4NA counts to 30, and orders, reads and represents numbers in the range 0 to 20

9

At the Farm

✡ Colour 2 horses brown.

Colour 4 pigs pink.

Colour 1 sheep purple.

Colour 5 cows black.

Colour 1 fish orange.

Colour 3 hens red.

Colour 4 ducks yellow.

Colour 2 birds blue.

STUDENT ASSESSMENT

✿ Look at the bowl below.

How many ⬭ ? _____

How many 🍌 ? _____

✿ Draw 4 🍓 and 2 🍎 and I 🍐 in the bowl.

✿ Colour 3 of the ⬭ green.

✿ Colour 2 of the 🍌 yellow.

✿ What other fruit can you put in the bowl?

Draw the fruit.

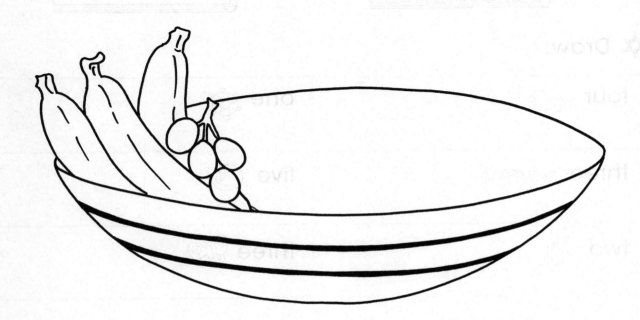

Unit
2
Counting to 5 (TRB pp. 26–29)
Whole numbers MAe-4NA counts to 30, and orders, reads and represents numbers in the range 0 to 20

11

Number Words

DATE:

✡ Complete the titles of the books by writing the number word.

✡ Draw:

four 	one 
three 	five 
two 	three 

All About Four

✵ Colour the groups that have **four** things.

How many groups did you colour? _____

✵ Can you find more groups of four in your classroom?

Draw them.

Unit 3 Groups of Things (TRB pp. 30–33)
Whole numbers MAe-4NA counts to 30, and orders, reads and represents numbers in the range 0 to 20

13

Less and More

Draw a group that is less.		Draw a group that is more.
	✏️ ✏️ ✏️ ✏️	
	✂️ ✂️ ✂️	
	🎒 🎒	
	📘 📘 📘	
	🧍 🧍 🧍 🧍	

Now draw your own group in the middle.

Draw a group that is less.		Draw a group that is more.

DATE:

STUDENT ASSESSMENT

✿ Finish the table.

1	one	
4		
	two	
		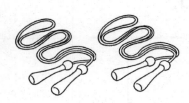
5		

✿ Colour the groups that have the **same** number of things.

✿ Look at the fish.

Draw a group that is **more**.

 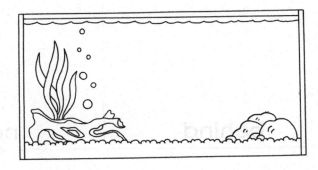

Unit 3 **Groups of Things** (TRB pp. 30–33)
Whole numbers MAe-4NA counts to 30, and orders, reads and represents numbers in the range 0 to 20

15

Where Am I?

✿ Write a word to match each picture.

near	between	next to	in front of	behind	under

_____   _____   _____

✿ Draw a picture for each word.

behind	near	in front of

Where Did Rosie Walk?

✿ Show where Rosie walked.

✿ Use the words to label how she walked.

past   around   under   across   over   through

Monkey Trouble

Draw things for Baby Monkey to go near, across, between, under and next to on her way back to Mother.

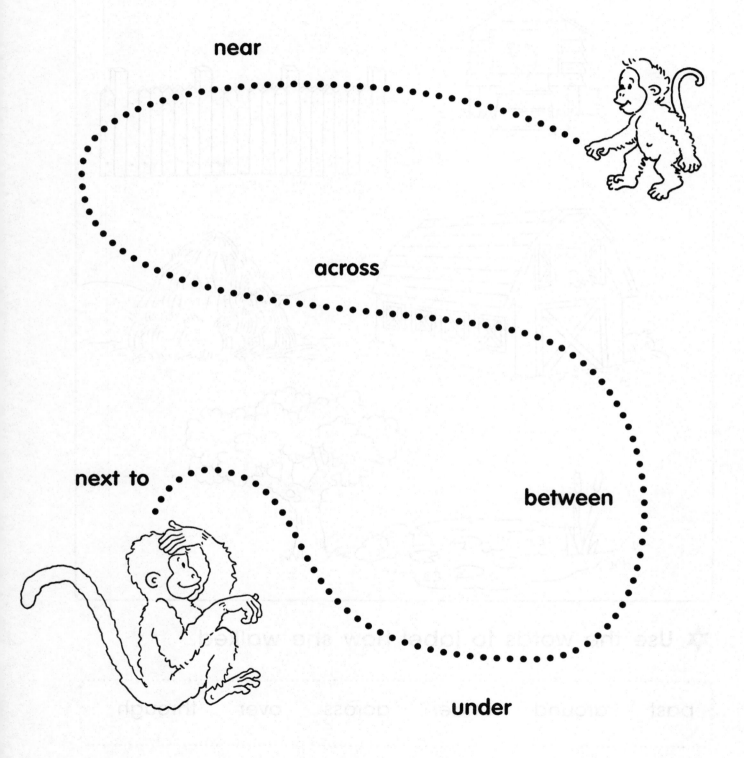

near

across

next to

between

under

Position (TRB pp. 34–37)
Position MAe-16MG describes position and gives and follows simple directions using everyday language

✿ Draw more farm animals in the correct places.

A sheep is **between** the horse and the cow.

A pig is **next to** the horse.

2 ducks are **near** the cow.

✿ Draw a farmer and the path she takes.

First, she walks **towards** the pig.

Then she walks **between** the horse
and the sheep.

Then the farmer walks **forwards** to the ducks.

At last she stands **next to** the cow.

Unit
4 **Position** (TRB pp. 34–37)
Position MAe-16MG describes position and gives and follows simple directions using everyday language

19

Rocket Blast-Off!

✶ Finish the poem. Use the numbers 5, 4, 3, 2, 1, 0.

Zoom! Zoom! Zoom!
We're going to the moon
Zoom! Zoom! Zoom!
We're going to the moon
We'll climb aboard a rocket ship
And go upon a little trip
Zoom! Zoom! Zoom!
We're going to the moon

_____, _____, _____, _____, _____, _____, Blast-off!

✶ Play a game.

You will need:

- a dice with the 6 changed to a 0
- a partner

How to play:

- In turn, roll the dice.
- Write the number you rolled on
  your rocket. Make sure
  it is in the correct order.
- When you have filled in 5, 4, 3, 2, 1, 0,
  colour in the "Blast-off".
- The first player to colour
  their "Blast-off" wins.

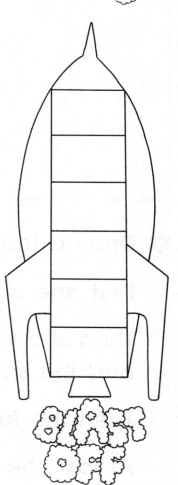

Before

☆ Write the number that comes **before**:

_____ 6 _____ 3 _____ 1 _____ 4 _____ 5 _____ 2

☆ Play a game.

You will need:

- some counters, a dice and a partner

How to play:

- Put your counters on START and roll the dice.
- Move to the number that comes **before** the number you rolled.
- The first player to pass FINISH wins.

Unit 5 **More Counting** (TRB pp. 38–41)
Whole numbers MAe-4NA counts to 30, and orders, reads and represents numbers in the range 0 to 20

21

Missing Numbers

✿ Write the missing numbers.

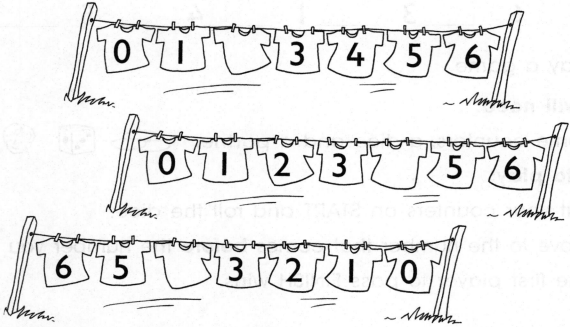

0 1 __ 3 4 5 6

0 1 2 3 __ 5 6

6 5 __ 3 2 1 0

✿ Write your own numbers.

✿ Write the missing numbers.

0
1

6 7 8 9

DATE:

STUDENT ASSESSMENT

✷ Complete the table.

Write the number that comes **before**.		Write the number that comes **after**.
	2	
	4	
	1	
	5	
	3	

✷ Count backwards from 6.

6 _____

✷ Fill in the missing numbers.

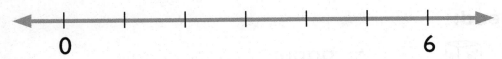

0   6

✷ Some shirts are missing. Draw them.

1   2   4   6

Unit 5 **More Counting** (TRB pp. 38–41)
Whole numbers MAe-4NA counts to 30, and orders, reads and represents numbers in the range 0 to 20

23

Dot Plates

✤ Show 5 in different ways. Use a paper plate and some counters.

Draw some of the ways you found.

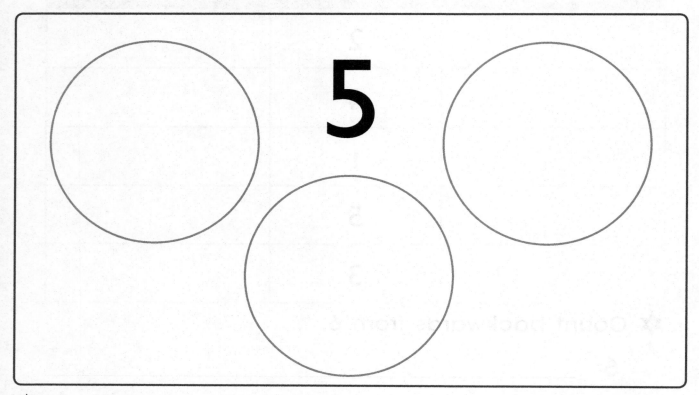

✤ Play a game.

You will need:

- a dice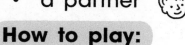
- a partner
- paper

How to play:

- In turn, roll a dice.
- Each time you roll 5, cover one of the paper plates above with a piece of paper.
- The first player to cover all their plates wins.

Glub! Glub!

☀ Draw dots on each frog to match the number of bugs it will eat. One has been done.

☀ On the lily pad, write the matching number word. One has been done.

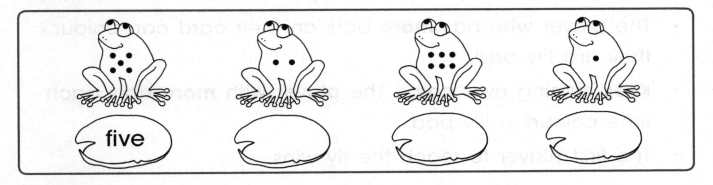

☀ Draw a line matching each frog to a lily pad and a bug. One has been done.

Unit 6 **Dot Patterns** (TRB pp. 42–45)
Whole numbers MAe-4NA counts to 30, and orders, reads and represents numbers in the range 0 to 20

25

Hungry Frogs

Play a game.

You will need:

- a partner
- one set of cards for each player
  (BLM 15 'Dot Patterns')

How to play:

- Choose a frog. Turn all your cards over and place them in a pile.

- Take a card from your pile. Your partner needs to do this, too.

- The player who has **more** dots on their card can colour their first lily pad.

- Keep turning over cards. The player with **more** dots each time colours a lily pad.

- The first player to reach the fly wins.

STUDENT ASSESSMENT

✿ Help Tommy Turtle get to the letterbox.
Colour in the stepping stones that show 5.

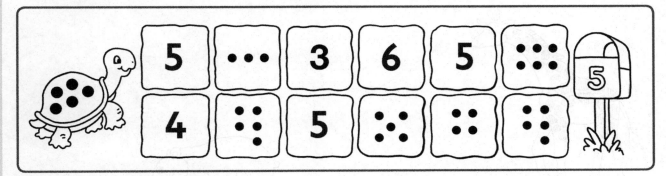

✿ Draw a friend for
Tommy Turtle who has
more dots than Tommy.

✿ Draw a friend for
Tommy Turtle who has
less dots than Tommy.

✿ Draw a friend with
the **same** number
of dots as Tommy Turtle.
Use a different pattern.

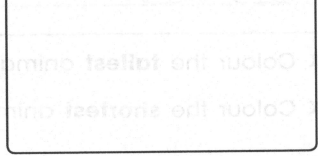

Unit 6 **Dot Patterns** (TRB pp. 42–45)
Whole numbers MAe-4NA counts to 30, and orders, reads and represents numbers in the range 0 to 20

27

At the Zoo

✿ Draw a home for each animal. Make sure each home is **taller than** the animal.

✿ Colour the **tallest** animal yellow.

✿ Colour the **shortest** animal blue.

Pencils

✦ Colour the **longest** pencil green.

✦ Look around the classroom. Draw something that is **longer than** the longest pencil.

✦ Colour the **shortest** pencil yellow.

✦ Look around the classroom. Draw something that is **shorter than** the shortest pencil.

✦ Draw a pencil that is **shorter than** the longest pencil but **longer than** the shortest pencil.

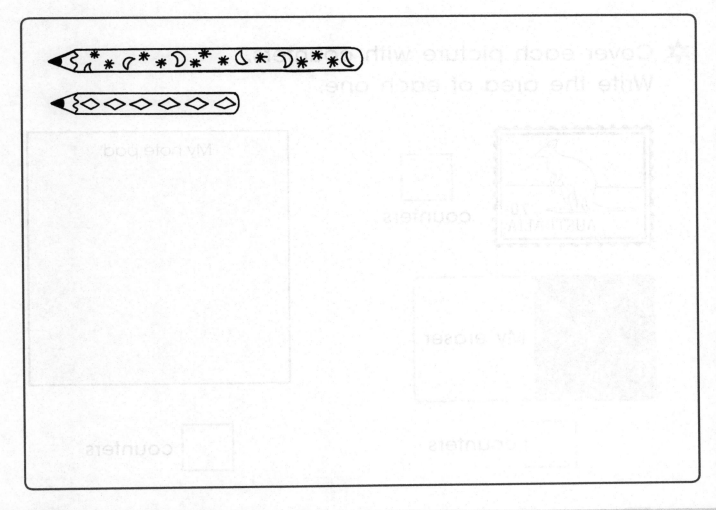

Unit **7** **Length and Area** (TRB pp. 46–49)
Length MAe-9MG describes and compares lengths and distances using everyday language

29

Area

Area is the whole surface
of something.

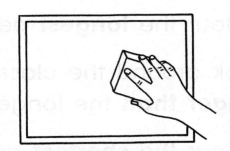

✸ Look at the shapes.

  ○ Circle the smallest area.

  ✓ Tick the largest area.

Colour the shapes with
the **same** area.

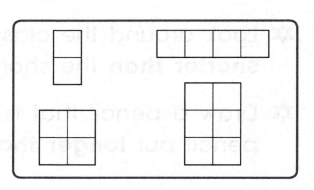

✸ Cover each picture with counters.
Write the area of each one.

□ counters

□ counters

STUDENT ASSESSMENT

✿ Look around the classroom. Draw something that is **longer than** your pencil.

✿ Colour the **longest** car red.

✿ How do you know which car is longer?

_____

_____

✿ Colour the shape with the largest area.

Unit 7   **Length and Area** (TRB pp. 46–49)
Length MAe-9MG describes and compares lengths and distances using everyday language
Area MAe-10MG describes and compares areas using everyday language

31

A Game of Numbers

Play a game.

You will need:

- a 10-sided dice
- a partner

How to play:

- In turn, roll the dice and trace over the number you rolled.
- The first person to trace over all their numbers wins.

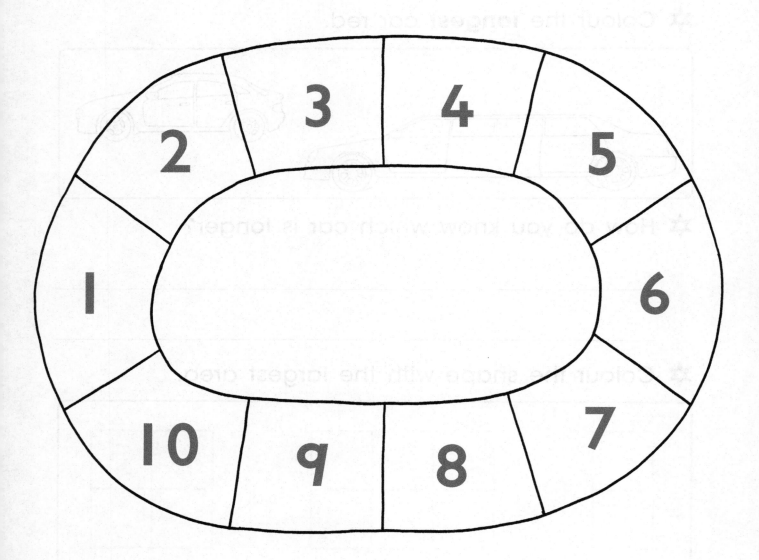

Unit 8 **Numbers to 10** (TRB pp. 50–53)
Whole numbers MAe-4NA counts to 30, and orders, reads and represents numbers in the range 0 to 20

Stepping Stones

✿ Colour the stepping stones to help the koala cross the river. You must colour **all** the stones from 1 to 10.

✿ Write the numbers in the correct order.

6   1   9   3   4   7   8   10   5   2

Unit 8   **Numbers to 10** (TRB pp. 50–53)
Whole numbers MAe-4NA counts to 30, and orders, reads and represents numbers in the range 0 to 20

33

In the Bush

DATE:

✿ Colour the 🦘 .
How many are there? ☐

✿ Colour the 🦢 .
How many are there? ☐

✿ Colour the 🐟 .
How many are there? ☐

✿ Draw 9 🐦 .

✿ Colour the 🐨 .
How many are there? ☐

✿ Draw 6 🐛 .

✿ Colour the 🦎 .
How many are there? ☐

34

Numbers to 10 (TRB pp. 50–53)
Whole numbers MAe-4NA counts to 30, and orders, reads and represents numbers in the range 0 to 20

DATE:

STUDENT ASSESSMENT

✿ Write the numbers from 1 to 10.

_____

✿ How many?

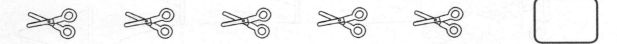

✿ Draw:

4 ✈

9 🎈

7 🪁

✿ Write the numbers in the correct order.

2   3   4   6   5   7   8   10   9

_____

Unit 8 **Numbers to 10** (TRB pp. 50–53)
Whole numbers MAe-4NA counts to 30, and orders, reads and represents numbers in the range 0 to 20

35

Are We There Yet?

✿ Count backwards.

10 _____ 0

✿ Write the numbers in the correct order to help the family get home.

✿ Play a game.

You will need: a 10-sided dice and a partner

• In turn, roll the dice until you roll a 10.

• Now colour in the ⚑.

• Keep rolling the dice until you get a 9. Colour in the ⚑.

• Keep rolling the dice until you have coloured all the signposts and the family is home ⚑.

Sorting Cards

You will need: a 10-sided dice

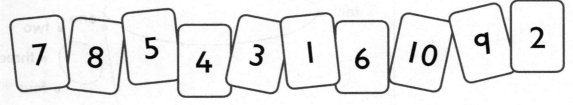

7   8   5   4   3   1   6   10   9   2

�֍ Sort the cards starting at 1.

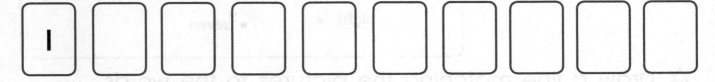

| 1 | | | | | | | | | |

✷ Sort again starting at 10.

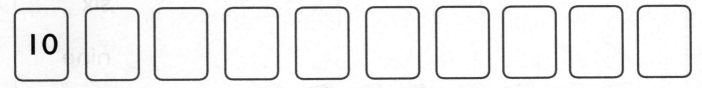

| 10 | | | | | | | | | |

✷ Throw the dice. Write the number in the box.

Now count **forwards** as far as you can and write the numbers.

[ ] _____

✷ Throw the dice. Write the number in the box.

Now count **backwards** as far as you can and write the numbers.

[ ] _____

Counting with Numbers to 10 (TRB pp. 54–57)
Whole numbers MAe-4NA counts to 30, and orders, reads and represents numbers in the range 0 to 20

37

A Different Dot-to-Dot

DATE:

✿ Join the dots.

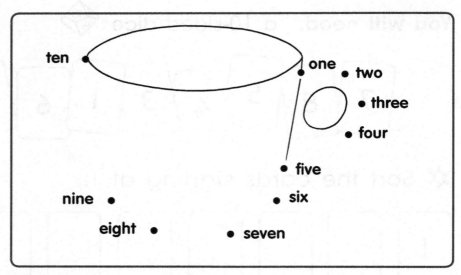

✿ Draw a line matching the pictures to the words.

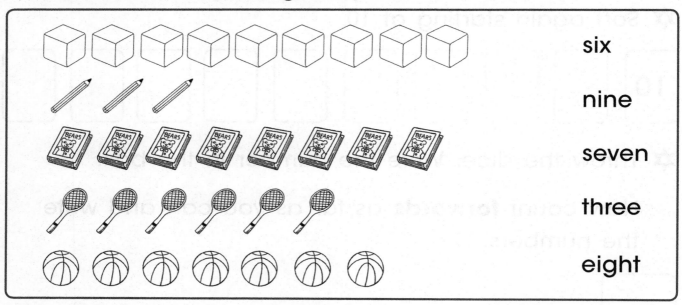

six

nine

seven

three

eight

✿ Draw a group of things for each number.

seven	four
ten	six
eight	five

STUDENT ASSESSMENT

DATE:

✿ Finish the counting patterns.

10   9   8   _____

5   _____

8   7   _____

✿ Start counting up from the number on the dice.

1   ___ ___ ___ ___ ___

5   ___ ___ ___ ___ ___

2   ___ ___ ___ ___ ___

✿ Draw a line to match the cards.

three	6
nine	7
six	3
seven	9

✿ Draw a group of things for each number.

five	seven
ten	eight

Unit 9   Counting with Numbers to 10 (TRB pp. 54–57)
Whole numbers MAe-4NA counts to 30, and orders, reads and represents numbers in the range 0 to 20

39

Using Ten Frames

✿ Show 8 on the ten frame.

How many empty spaces? _____

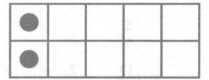

✿ How many dots? _____

How many empty spaces? _____

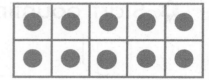

✿ How many dots? _____

How many empty spaces? _____

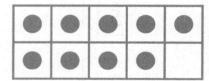

✿ How many dots? _____

How many empty spaces? _____

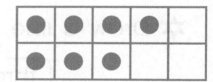

✿ How many dots? _____

How many empty spaces? _____

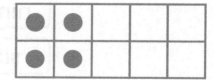

✿ How many dots? _____

How many empty spaces? _____

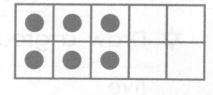

✿ How many dots? _____

How many empty spaces? _____

✿ Show 6 another way.

Ten Frame Trains

Complete the trains. One has been done.

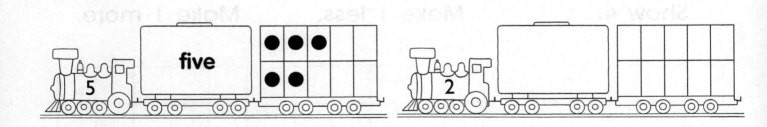

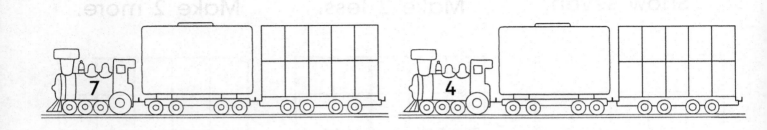

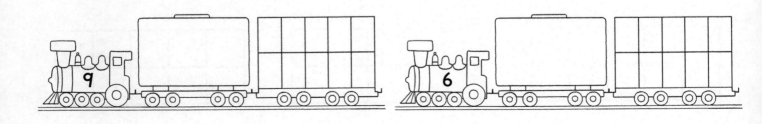

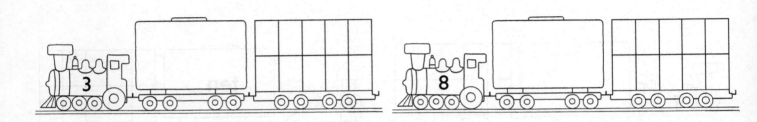

Unit **10** **Ten Frames** (TRB pp. 58–61)
Whole numbers MAe-4NA counts to 30, and orders, reads and represents numbers in the range 0 to 20

41

Some More, Some Less

✡ Complete the ten frames.

Show 4.

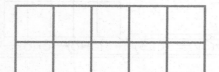

Make 1 less.

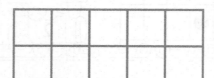

Make 1 more.

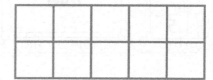

Show seven.

Make 2 less.

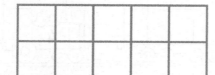

Make 2 more.

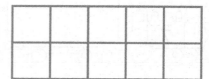

Show nine.

Make two less.

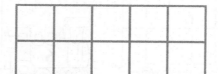

Make 1 more.

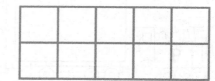

✡ Complete the trains.

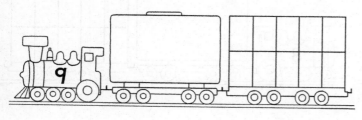

STUDENT ASSESSMENT

✹ Use a ten frame to show:

3	7	4

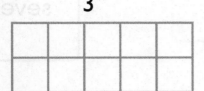 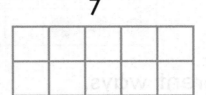 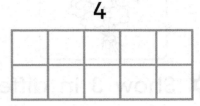

✹ What number can you see?

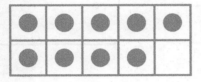 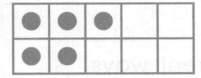 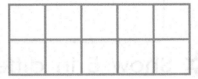

_____ _____ _____

✹ Show two **more** than: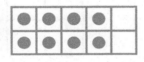

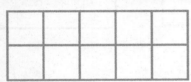

✹ Show one **less** than: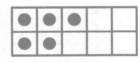

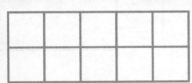

✹ Draw a line to match the numbers.

four

six

10

7

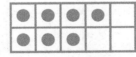

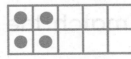

Unit
10
Ten Frames (TRB pp. 58–61)
Whole numbers MAe-4NA counts to 30, and orders, reads and represents numbers in the range 0 to 20

43

Different Ways

✴ Look at the different ways we can show 7.

 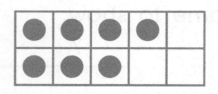 seven

✴ Show 3 in different ways.

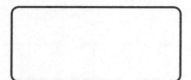  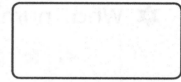

✴ Show 5 in different ways.

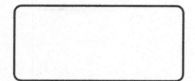 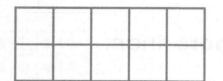 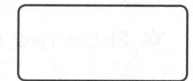

✴ Show 10 in different ways.

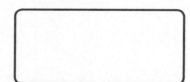 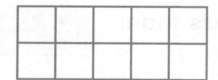 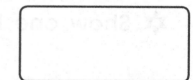

✴ Show 8 in different ways.

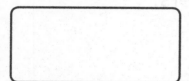 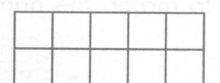 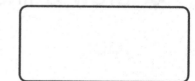

✴ Draw a line to match the numbers.

 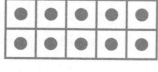

four

6   9   4   10

44   Unit 11   **Counting and Comparing Groups** (TRB pp. 62–65)
Whole numbers MAe-4NA counts to 30, and orders, reads and represents numbers in the range 0 to 20
Multiplication and division MAe-6NA groups, shares and counts collections of objects, describes using everyday language, and records using informal methods

Lots of Toys

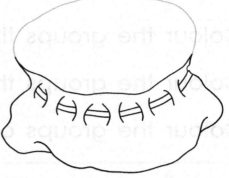

✿ How old are you? _____

✿ Draw that many toys
in the bag.

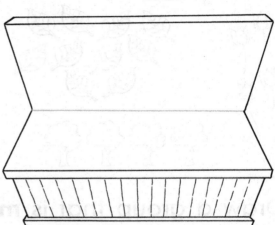

✿ Draw **more** toys than
you drew in the bag.

There are _____ toys.

✿ There are _____ books.

Draw **more** books.

Draw **less** books.

Unit 11 **Counting and Comparing Groups** (TRB pp. 62–65)
Whole numbers MAe-4NA counts to 30, and orders, reads and represents numbers in the range 0 to 20
Multiplication and division MAe-6NA groups, shares and counts collections of objects, describes using everyday language, and
records using informal methods

45

Colour the Groups

✿ Colour the groups that are **more than** 5 green.

✿ Colour the groups that are **less than** 5 blue.

✿ Colour the groups of 5 yellow.

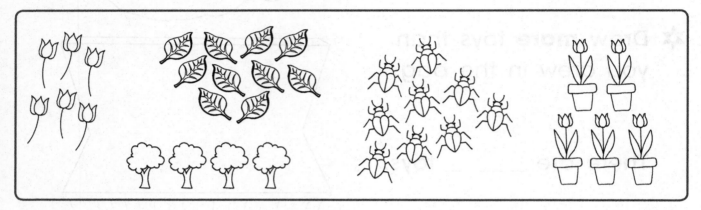

✿ Draw a group that is **more than** 5.

✿ Draw another group that is **more than** 5.

✿ Draw another group that is **more than** 5.

Unit **11**

Counting and Comparing Groups (TRB pp. 62–65)
Whole numbers MAe-4NA counts to 30, and orders, reads and represents numbers in the range 0 to 20
Multiplication and division MAe-6NA groups, shares and counts collections of objects, describes using everyday language, and records using informal methods

STUDENT ASSESSMENT

☆ Colour the groups that show 3.

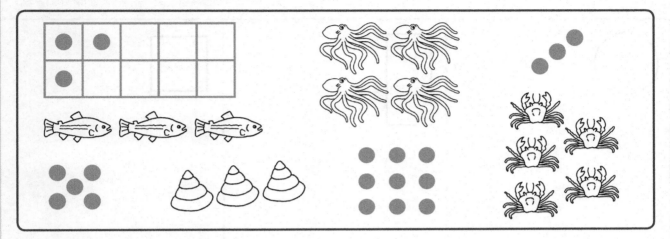

☆ Colour the group that has **less**.

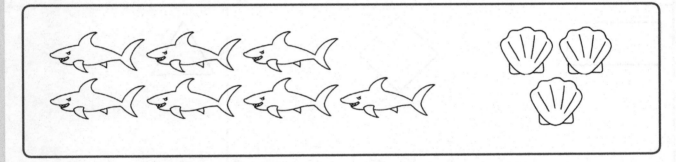

☆ Colour all the groups that are **more than** 5.

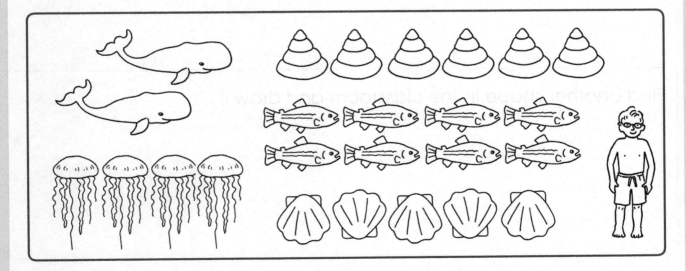

Unit
11
Counting and Comparing Groups (TRB pp. 62–65)
Whole numbers MAe-4NA counts to 30, and orders, reads and represents numbers in the range 0 to 20
Multiplication and division MAe-6NA groups, shares and counts collections of objects, describes using everyday language, and records using informal methods

47

Shapes You Can See

Look at each shape. Draw something from your classroom that has the same shape.

Find another shape in the classroom and draw it.

Unit 12 **2D Shapes** (TRB pp. 66–69)
Two-dimensional space MAe-15MG manipulates, sorts and describes representations of two-dimensional shapes, including circles, triangles, squares and rectangles, using everyday language

Find the Shape

✸ Look at the picture below.
   Colour all the rectangles blue.

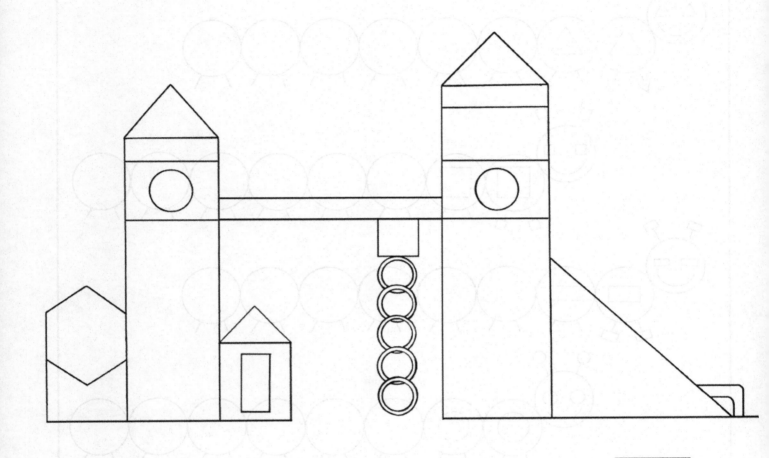

✸ How many rectangles are in the picture?

✸ What other shapes can you see?

_____

_____

Shape Caterpillars

✿ Finish the patterns on the caterpillars.

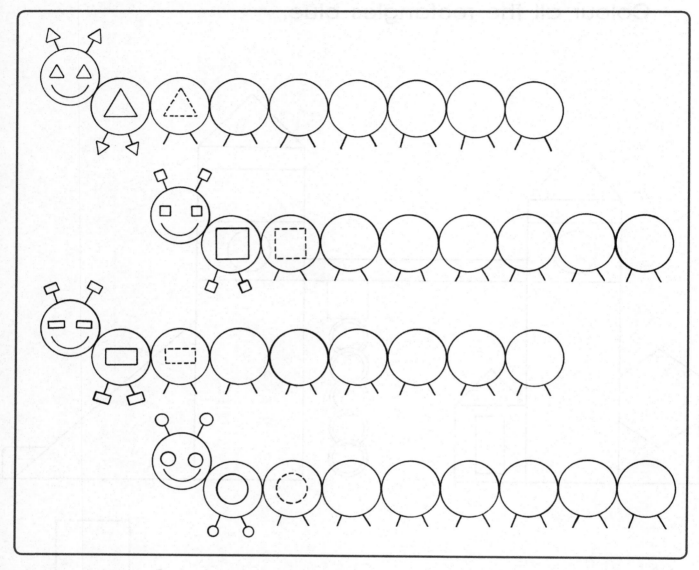

✿ Draw your own shapes on the caterpillar.

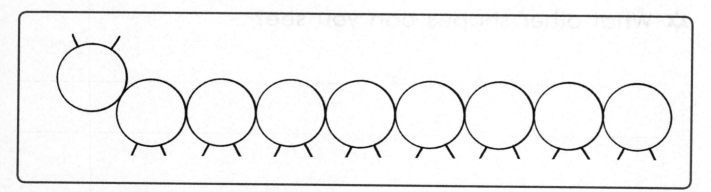

Unit 12 **2D Shapes** (TRB pp. 66–69)
Two-dimensional space MAe-15MG manipulates, sorts and describes representations of two-dimensional shapes, including circles, triangles, squares and rectangles, using everyday language

DATE:

STUDENT ASSESSMENT

✸ Look at all the shapes. Colour the triangles red.

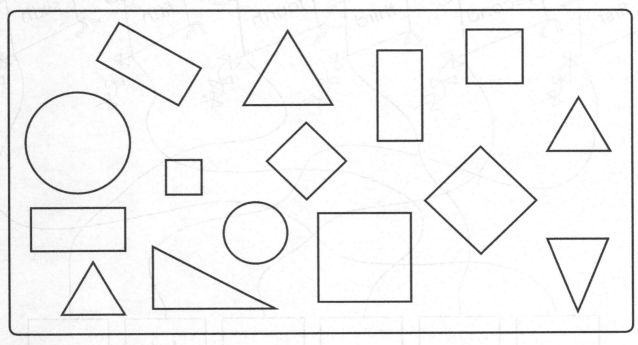

✸ This is a rectangle.

Draw something from your classroom that has a rectangle.

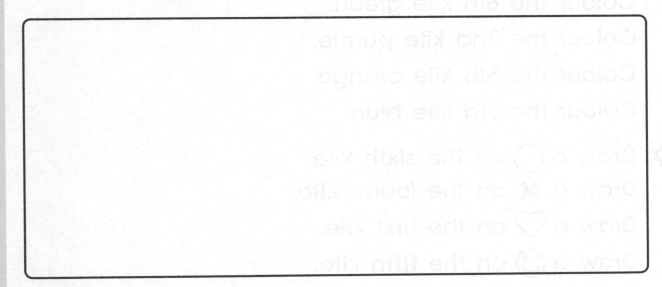

Unit
12
2D Shapes (TRB pp. 66–69)
Two-dimensional space MAe-15MG manipulates, sorts and describes representations of two-dimensional shapes, including circles, triangles, squares and rectangles, using everyday language

51

Colourful Kites

✿ Fill in the boxes below.

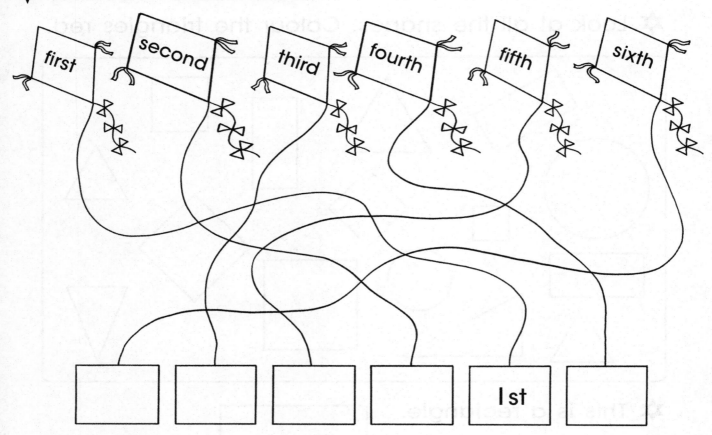

✿ Colour the 1st kite red.

Colour the 4th kite yellow.

Colour the 6th kite green.

Colour the 2nd kite purple.

Colour the 5th kite orange.

Colour the 3rd kite blue.

✿ Draw a ◯ on the sixth kite

Draw a ✖ on the fourth kite.

Draw a ♡ on the first kite.

Draw a ☺ on the fifth kite.

Clowns in Colour

✡ Colour three hats red.

Colour the other hats any colour you like.

Add more colour to your clowns.

✡ With a partner, take it in turn to describe your clowns.

✡ Now colour in the clowns below to match your partner's clowns.

Your partner will need to tell you which hats to colour red.

They will need to tell you all the other things to colour.

Unit **13**   **Ordinal Number** (TRB pp. 70–73)
Whole numbers MAe-4NA counts to 30, and orders, reads and represents numbers in the range 0 to 20

53

Ducks in a Row

✿ Draw a line to match the word with the duck.

One has been done.

fourth   seventh   first   fifth   sixth   third   second

✿ Colour the 3rd and 7th ducks blue.

Colour the 5th and 1st ducks red.

Colour the 4th duck green.

Colour the 6th and 2nd ducks yellow.

STUDENT ASSESSMENT

☆ Draw a ✖ on the first car.

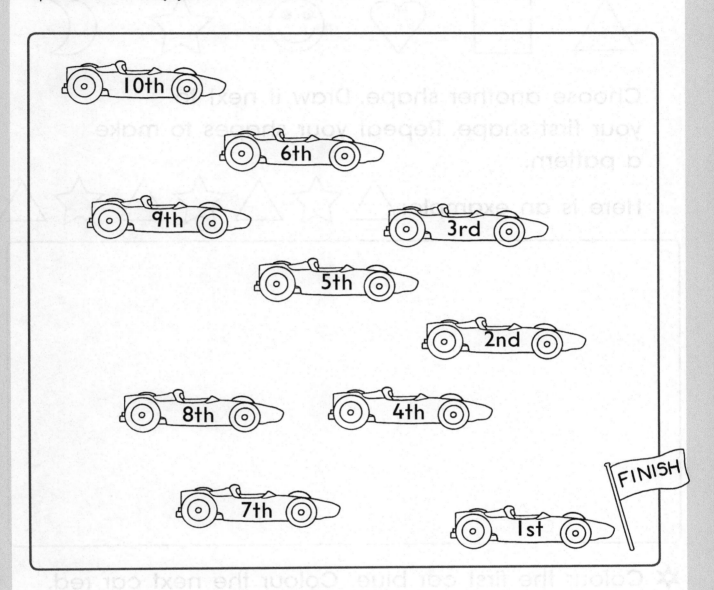

☆ Colour the **fourth** car red.

Colour the **seventh** car green.

Colour the **second** car yellow.

Colour the **fifth** car blue.

Make a Pattern

☼ Choose one shape and draw it in the box.

Choose another shape. Draw it next to your first shape. Repeat your shapes to make a pattern.

Here is an example:

☼ Colour the first car blue. Colour the next car red. Repeat your colouring to make a pattern.

☼ Draw your own pattern on another sheet of paper.

Patterns (TRB pp. 74–77)
Patterns and algebra MAe-8NA recognises, describes and continues repeating patterns

Musical Patterns

✿ Look at the pattern. Make the pattern by clapping.

Draw the same pattern.

✿ Make another pattern using  and 🪇 .

Clap this pattern.

✿ Finish the pattern below.

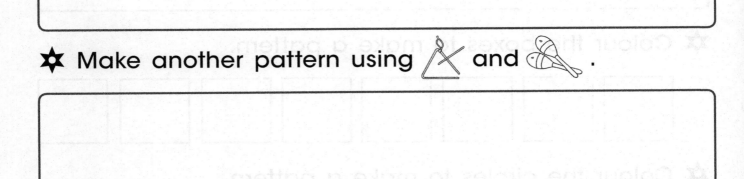

✿ Make a musical pattern using some of the instruments above.

Unit
14
Patterns (TRB pp. 74–77)
Patterns and algebra MAe-8NA recognises, describes and continues repeating patterns

57

Shapes and Patterns

✿ Copy the pattern.

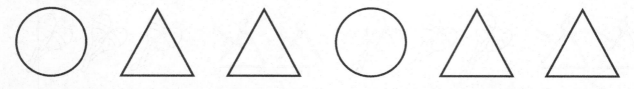

✿ Colour the boxes to make a pattern.

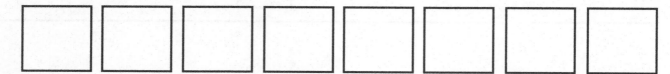

✿ Colour the circles to make a pattern.

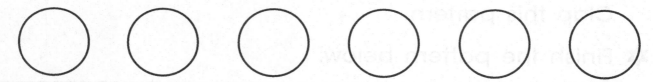

✿ Finish the patterns.

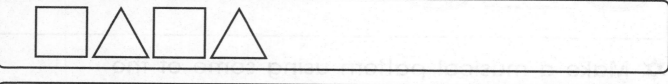

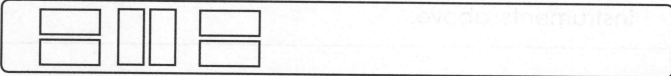

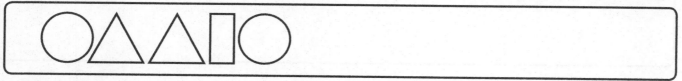

STUDENT ASSESSMENT

✿ **Colour to make a pattern.**

✿ **Copy the pattern.**

✿ **Finish the patterns.**

✿ **Draw your own pattern.**

Unit 14 **Patterns** (TRB pp. 74–77)
Patterns and algebra MAe-8NA recognises, describes and continues repeating patterns

59

My Timeline

✿ Draw yourself when you were a baby.

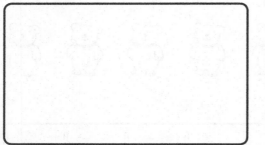

I liked _____

_____

✿ Draw yourself today.

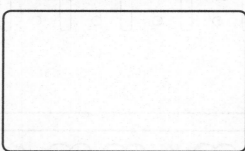

I like _____

_____

✿ Draw what you think you will look like in high school.

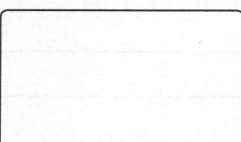

I think I will be _____

_____

✿ Draw what you think you will look like when you are an adult.

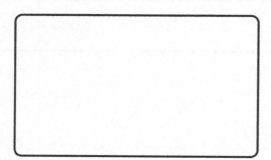

I think I will be _____

_____

My School Day

I wake up.

Then I _____

_____

_____

I come to school by

_____

_____

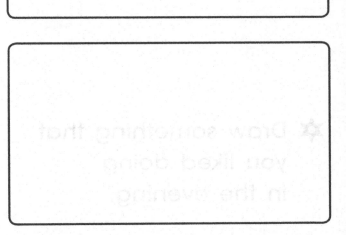

At school, I like to

_____

_____

Yesterday

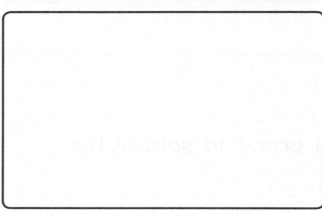

✿ **Think about yesterday.**

Draw something that
you liked doing
in the morning.

✿ Draw something that
you liked doing
in the afternoon.

✿ Draw something that
you liked doing
in the evening.

✿ What was your favourite time of the day?

_____

Why? _____

STUDENT ASSESSMENT

✿ **Look at the pictures. There is a 1 under the first thing the boy does.**

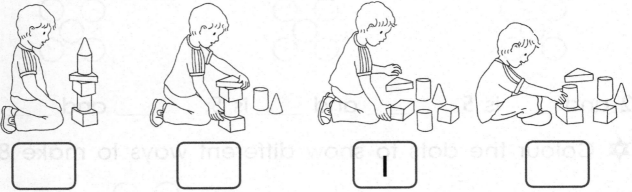

| | | 1 | |

Write 2 under what he does next.

Write 3 under what he does next.

Write 4 under the last thing he does.

✿ **Draw something you do:**

- after school.
- before school.

✿ **Draw something you do at these times:**

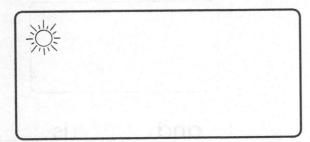

Unit 15   **Time** (TRB pp. 78–81)
Time MAe-13MG sequences events, uses everyday language to describe the durations of events, and reads hour time on clocks

63

Colour the Dots

✿ Colour 2 dots blue and the others red.

✿ Colour the dots to show different ways to make 5.

     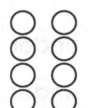

2 and ____ is 5.   ____ and ____ is 5.   ____ and ____ is 5.

✿ Colour the dots to show different ways to make 8.

_____ and _____ is 8.   _____ and _____ is 8.

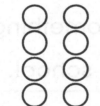

_____ and _____ is 8.   _____ and _____ is 8.

✿ Draw dots for another number in each box.
Colour the dots to show different ways to make
the number.

_____ and _____ is _____   _____ and _____ is _____

Lovely Ladybirds

✺ Look at the ladybird. Write about her dots.

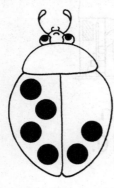

_____ and _____ is _____

✺ Draw different dots on each side of the ladybirds to show the same number.

_____ and _____ is _____   _____ and _____ is _____

✺ Show different ways to make 9.

Write about what you have drawn.

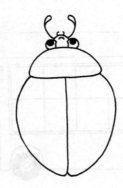   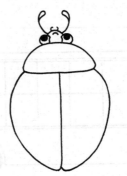

_____   _____

Unit
16   **Understanding Numbers to 10** (TRB pp. 82–85)
Whole numbers MAe-4NA counts to 30, and orders, reads and represents numbers in the range 0 to 20

65

Ten Bus

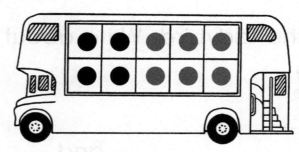

4 and 6 is 10.

✺ Draw red dots and blue dots on each bus to show different ways to make 10.

✺ Write a number story for each bus.

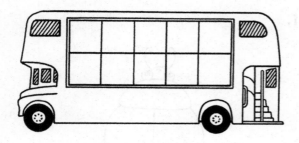

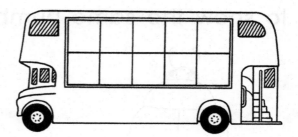

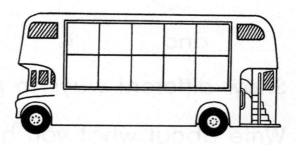

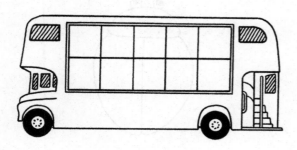

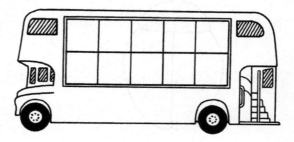

STUDENT ASSESSMENT

✿ There are 6 puppies in a litter. Some are brown and some are black. Colour the puppies to show what the litter might look like.

✿ How else might the litter look?

✿ 10 balls are in a box. Some are blue and some are red. Colour the balls to show one way they might look.

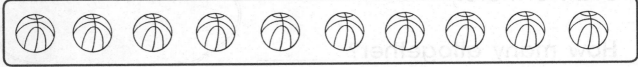

✿ There are 7 eggs in a nest. Some are white and some are brown. Draw one way they might look.

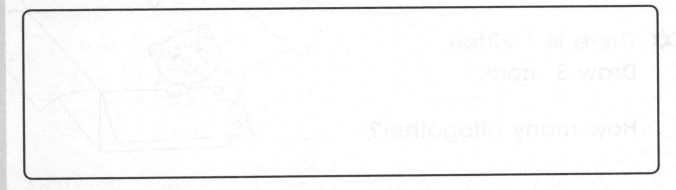

Unit
16
Understanding Numbers to 10 (TRB pp. 82–85)
Whole numbers MAe-4NA counts to 30, and orders, reads and represents numbers in the range 0 to 20

67

Lots of Pets

✹ There are 4 fish.
Draw 3 more.

How many altogether? _____

✹ There are 6 puppies.
Draw 2 more.

How many altogether? _____

✹ There are 2 ducks.
Draw 5 more.

How many altogether? _____

✹ There is 1 kitten.
Draw 3 more.

How many altogether? _____

68 **Unit 17** **Beginning Addition** (TRB pp. 86–89)
Addition and subtraction MAe-5NA combines, separates and compares collections of objects, describes using everyday
language, and records using informal methods

In the Paddocks

✿ Draw 3 sheep and draw 3 sheep.

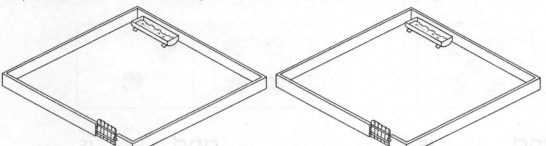

How many altogether?

_____

✿ Draw 3 cows and draw 6 cows.

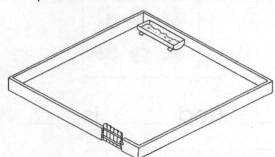 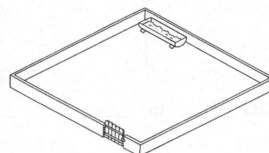

How many altogether?

_____

✿ Draw 4 horses and draw 3 horses.

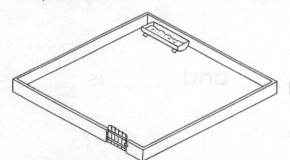 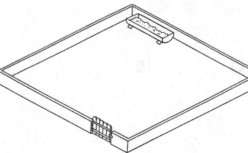

How many altogether?

_____

✿ Write your own story.

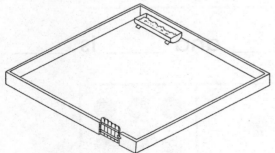 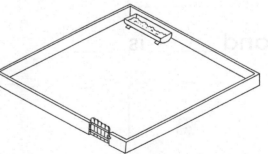

There are _____ and _____ .

There are _____ altogether.

Unit 17 **Beginning Addition** (TRB pp. 86–89)
Addition and subtraction MAe-5NA combines, separates and compares collections of objects, describes using everyday language, and records using informal methods

69

Dots on Dominoes

Fill in the numbers.

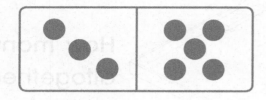

_____ and _____ is _____

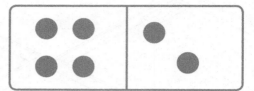

_____ and _____ is _____

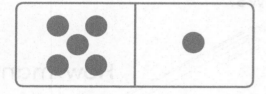

_____ and _____ is _____

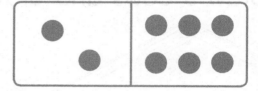

_____ and _____ is _____

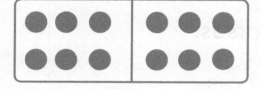

_____ and _____ is _____

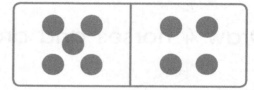

_____ and _____ is _____

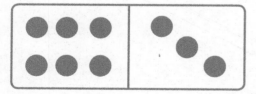

_____ and _____ is _____

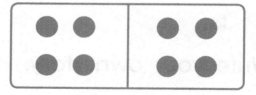

_____ and _____ is _____

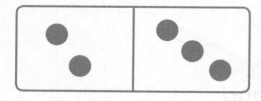

_____ and _____ is _____

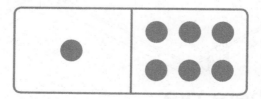

_____ and _____ is _____

70 Unit 17 **Beginning Addition** (TRB pp. 86–89)
Addition and subtraction MAe-5NA combines, separates and compares collections of objects, describes using everyday
language, and records using informal methods

STUDENT ASSESSMENT

There are _____ books and _____ books.

There are _____ books altogether.

✿ Draw 5 more pencils.

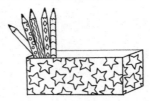

There are _____ pencils altogether.

✿ Look at the shells. Write a story about them.

_____

✿ Look at the domino.

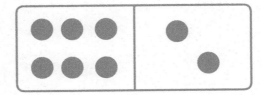

How many dots altogether? _____

How did you work that out? _____

_____

Unit
17
Beginning Addition (TRB pp. 86–89)
Addition and subtraction MAe-5NA combines, separates and compares collections of objects, describes using everyday language, and records using informal methods

71

Sorting Objects

You will need:

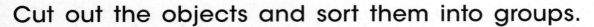

- a copy of **BLM 41 'Shape of Things'**
- scissors, glue, paper

Cut out the objects and sort them into groups.

Choose one group of objects to paste below.

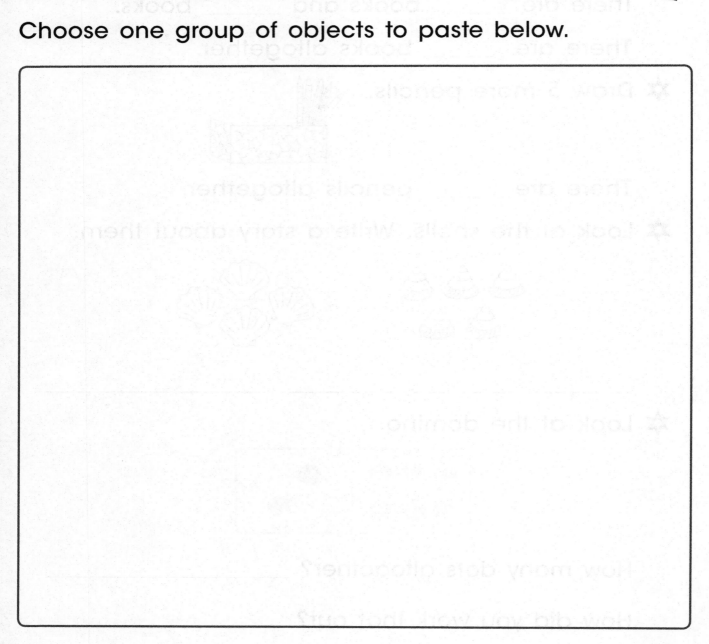

Paste your other groups on another sheet of paper.

What Could It Be?

Look around. Draw things that are the same as these objects.

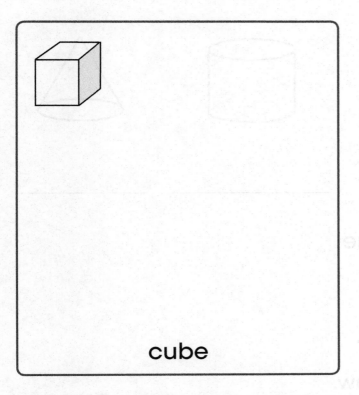

cube

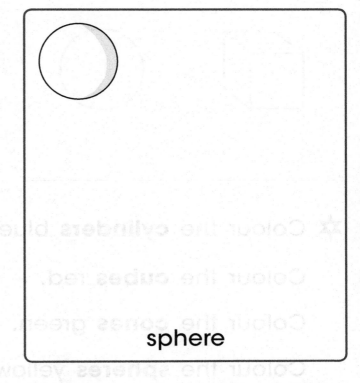

sphere

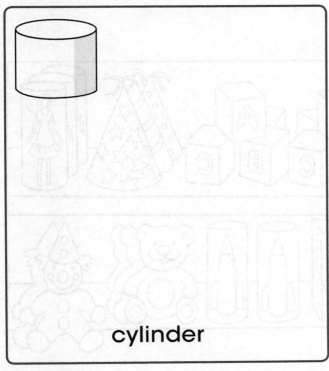

cylinder

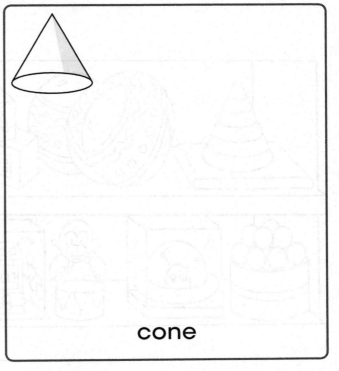

cone

Unit **18** **More About Shapes and Objects** (TRB pp. 90–93)
Three-dimensional space MAe-14MG manipulates, sorts and represents three-dimensional objects and describes them using everyday language

73

Can You Find the Objects?

✸ Write the name of each object.

_____ _____ _____ _____

✸ Colour the **cylinders** blue.

Colour the **cubes** red.

Colour the **cones** green.

Colour the **spheres** yellow.

DATE:

STUDENT ASSESSMENT

✦ Write the name of each object.

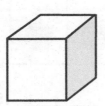

_____

_____

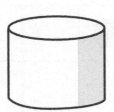

_____

_____

✦ Draw something that is the same as each object.

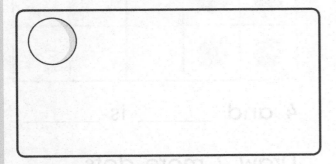

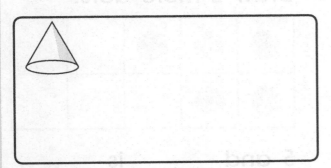

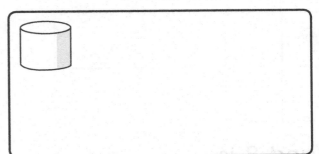

Unit
18
More About Shapes and Objects (TRB pp. 90–93)
Three-dimensional space MAe-14MG manipulates, sorts and represents three-dimensional objects and describes them using everyday language

75

Addition with Ten Frames

Draw 5 more dots.

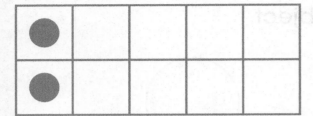

2 and 5 is _____

Draw 2 more dots.

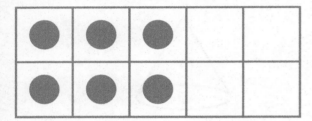

6 and 2 is _____

Draw 4 more dots.

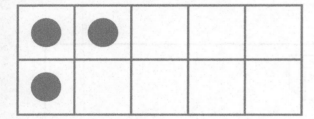

3 and 4 is _____

Draw 9 more dots.

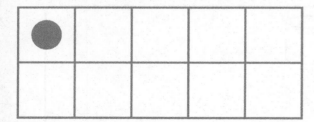

1 and 9 is _____

Draw 2 more dots.

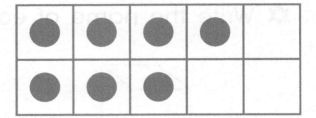

7 and _____ is _____

Draw 5 more dots.

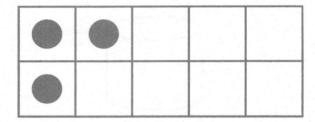

3 and _____ is _____

Draw 1 more dot.

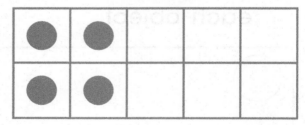

4 and _____ is _____

Draw 2 more dots.

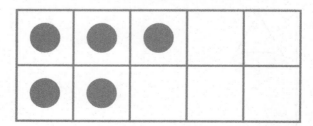

5 and _____ is _____

Unit 19 **More About Addition** (TRB pp. 94–97)
Addition and subtraction MAe-5NA combines, separates and compares collections of objects, describes using everyday language, and records using informal methods

Dominoes and Dice

✿ Draw 4 and **one** more.

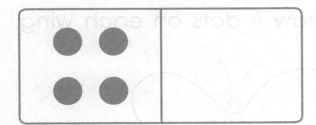

Draw 1 and **one** more.

Draw 6 and **one** more.

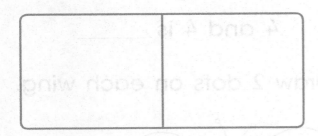

Draw 5 and **one** more.

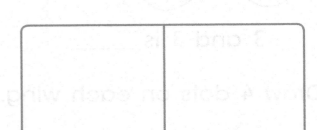

Draw 3 and **one** more.

Draw 2 and **one** more.

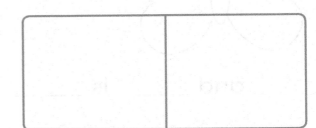

✿ Draw **one** more than:

Draw **one** more than: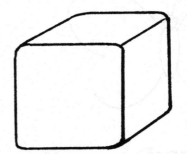

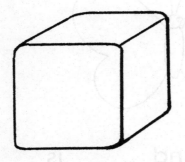

Unit 19 **More About Addition** (TRB pp. 94–97)
Addition and subtraction MAe-5NA combines, separates and compares collections of objects, describes using everyday language, and records using informal methods

77

Beautiful Butterflies

Butterflies have the same pattern on both wings.

Draw 3 dots on each wing.   Draw 4 dots on each wing.

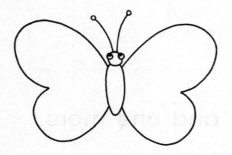   

3 and 3 is _____   4 and 4 is _____

Draw 4 dots on each wing.   Draw 2 dots on each wing.

_____ and _____ is _____   _____ and _____ is _____

Draw dots on these butterflies. Write a number sentence.

_____ and _____ is _____   _____ and _____ is _____

STUDENT ASSESSMENT

DATE: _____

☆ Use the ten frames to work out the answers.

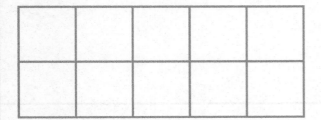

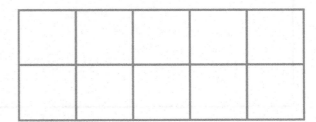

4 and 5 is _____   2 and 8 is _____

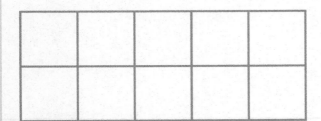

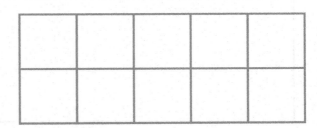

6 and 3 is _____   3 and 2 is _____

☆ I more than 7 is _____   Double 3 is _____

I more than 4 is _____   Double I is _____

I more than 3 is _____   Double 5 is _____

I more than 2 is _____   Double 6 is _____

I more than 8 is _____   Double 2 is _____

I more than 6 is _____   Double 4 is _____

Unit 19
More About Addition (TRB pp. 94–97)
Addition and subtraction MAe-5NA combines, separates and compares collections of objects, describes using everyday language, and records using informal methods

79

Drawing Groups

✿ Draw 3 groups of 4. 🍓 🍓 🍓 🍓 How many strawberries?

```
┌─────────────────────────────────────────┐
│                                         │
│                                         │
│                                         │
│                                         │
└─────────────────────────────────────────┘
```

✶ Draw 1 group of 7. 🌷🌷🌷🌷🌷🌷🌷 How many flowers?

```
┌─────────────────────────────────────────┐
│                                         │
│                                         │
│                                         │
│                                         │
└─────────────────────────────────────────┘
```

✶ Draw 4 groups of 5. 🦋 🦋 🦋 🦋 🦋 How many butterflies?

```
┌─────────────────────────────────────────┐
│                                         │
│                                         │
│                                         │
│                                         │
└─────────────────────────────────────────┘
```

✿ Draw 5 groups of 2 flowers. ❀ ❀ How many flowers?

```
┌─────────────────────────────────────────┐
│                                         │
│                                         │
│                                         │
└─────────────────────────────────────────┘
```

Equal Groups

✹ **How many groups of 2?**

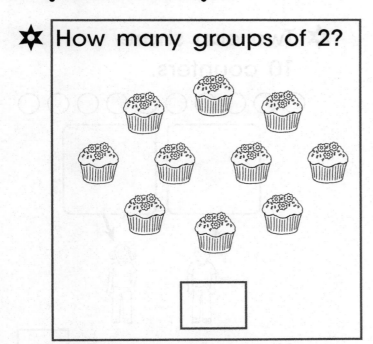

How many groups of 3?

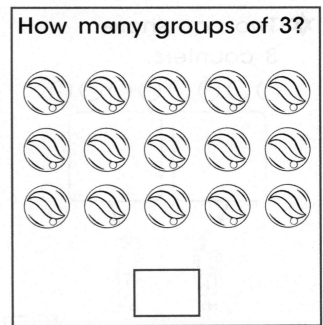

✹ **How many groups of 3?**

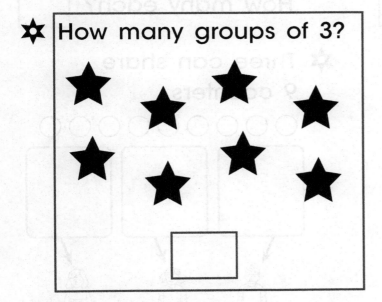

How many groups of 6?

✹ **How many groups of 3 balloons can you make from 12 balloons?**

Unit 20 **Grouping and Sharing** (TRB pp. 98–101)
Multiplication and division MAe-6NA groups, shares and counts collections of objects, describes using everyday language, and records using informal methods

81

Sharing

✸ Two can share
8 counters.

How many each?

✸ Two can share
10 counters.

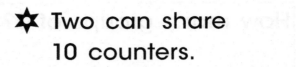

How many each?

✸ Three can share
6 counters.

How many each?

✸ Three can share
9 counters.

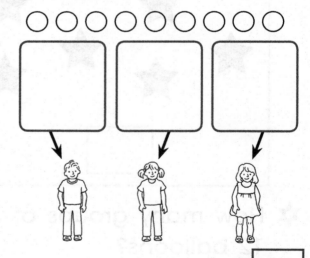

How many each?

✸ Show how you can
share 12 counters.

Unit 20 **Grouping and Sharing** (TRB pp. 98–101)
Multiplication and division MAe-6NA groups, shares and counts collections of objects, describes using everyday language, and records using informal methods

Unit
20 STUDENT ASSESSMENT

✿ Draw 3 groups of 5. ❀❀❀❀❀ How many flowers?

✿ Two can share 14 counters.

○○○○○○○○○○○○○○

How many each? ☐

✿ Circle groups of 2.

How many
groups of 2? ☐

✿ Show how you can share 8 counters.

Unit
20 Grouping and Sharing (TRB pp. 98–101)
Multiplication and division MAe-6NA groups, shares and counts collections of objects, describes using everyday
language, and records using informal methods

83

Is It Heavy or Light?

✵ Pick up these things. Are they **heavy** or **light**?

✵ Draw them in the correct box.

Heavy	Light

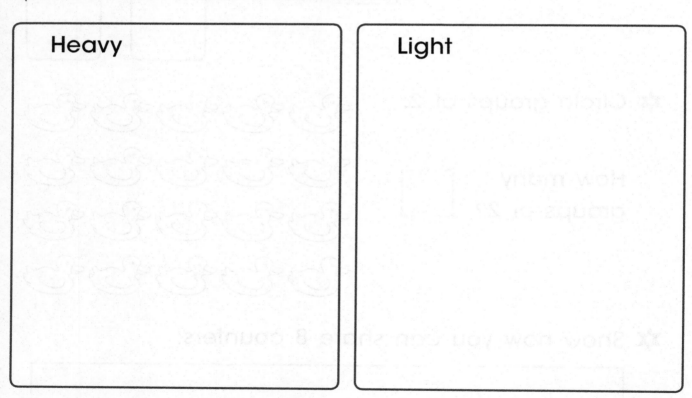

✵ On another sheet of paper, draw some other things in your classroom that are **heavy** or **light**.

Heavier Than a Book

✸ Pick up a book.

Now find 5 things that are **heavier** than your book. Draw them.

✸ What was the **heaviest** thing you found?

_____

✸ What was the **lightest** thing you found?

_____

Unit **21** **Mass** (TRB pp. 102–105)
Mass MAe-12MG describes and compares the masses of objects using everyday language

85

Sorting Toys

✿ Draw a toy in the top box.

Draw a toy that is **heavier** than your toy.

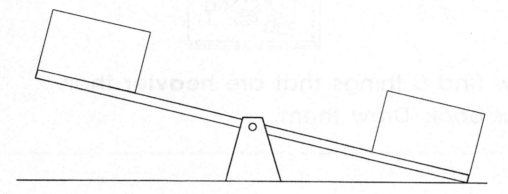

✿ Draw your toy in the bottom box.

Draw a toy that is **lighter** than your toy.

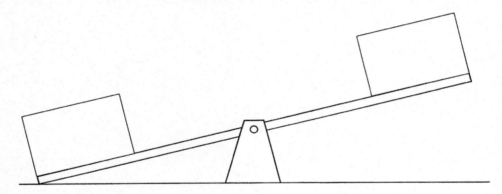

✿ Draw all your toys from heaviest to lightest.

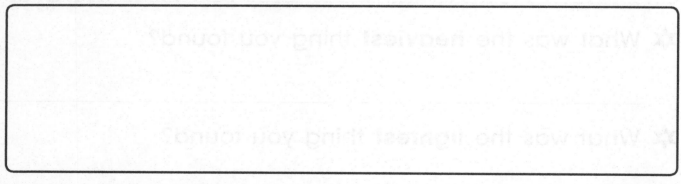

heavy ⟷ light

STUDENT ASSESSMENT

✿ Draw something that is **heavy**.

✿ Draw something that is **light**.

✿ Draw something that is **heavier** than a 🐻.

✿ Colour the **lightest** toy.

Number Search

✿ Lightly colour over the numbers.

11 is red, 12 is green, 13 is purple, 14 is light blue,
15 is dark blue, 16 is orange, 17 is yellow,
18 is brown, 19 is pink and 20 is grey.

11	12	13	14	15	16	17
18	19	20	16	18	17	19
20	14	11	13	15		19
18	17	17	15	14		13
12	11	19	17	15	13	11
18	16	14	12	13	11	16
11	19	14	17	11	12	15	18
11	11	15	12	16	13	17
14	19	18	12	13	14
15	16	17	18	19	14	17

✿ Write how many of each number you found.

11	12	13	14	15	16	17	18	19	20

Building Climb

✡ Write a number on each floor of this building. The first two have been done for you.

✡ Play this game with a partner to see who can climb their building first.

You will need:

- a dice
- counters

How to play:

- Roll a dice 4 times and move that many spaces.
- Now your partner should do the same.
- If you have made it to the top, score 1 point. Otherwise the person on the highest floor scores 1 point.
- The first person with 10 points wins.

2

1

Unit
22
Numbers Beyond 10 (TRB pp. 106–109)
Whole numbers MAe-4NA counts to 30, and orders, reads and represents numbers in the range 0 to 20

89

Number Challenge

✸ Complete the counting sequence in each row.

12	13	14	15	16	17	18
			17			
			12			
			16			
			14			
			13			
			18			

✸ Find a path from 20 to a number at the bottom that includes all the numbers from 11 to 20.

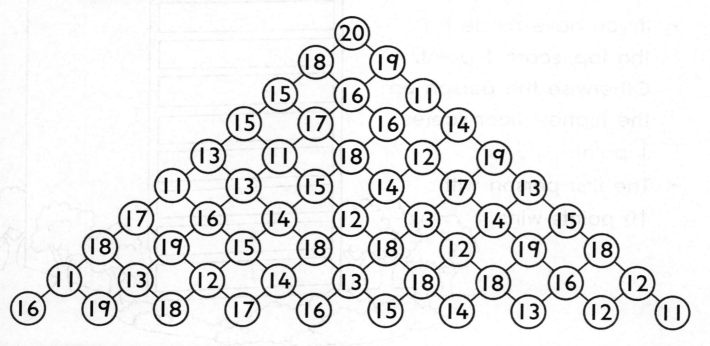

Numbers Beyond 10 (TRB pp. 106–109)
Whole numbers MAe-4NA counts to 30, and orders, reads and represents numbers in the range 0 to 20

DATE:

STUDENT ASSESSMENT

✹ Fill in the missing numbers.

1	2	3			6		8		10
11			14			17			20

✹ Write the numbers in the correct order.

17   15   18   16

_____

12   14   13   11

_____

10   12   9   11

_____

20   18   19   17

_____

✹ Write the five numbers that come **after** 10 when you are counting.

_____

Unit 22 **Numbers Beyond 10** (TRB pp. 106–109)
Whole numbers MAe-4NA counts to 30, and orders, reads and represents numbers in the range 0 to 20

91

Finding a Half

✸ Colour a half.

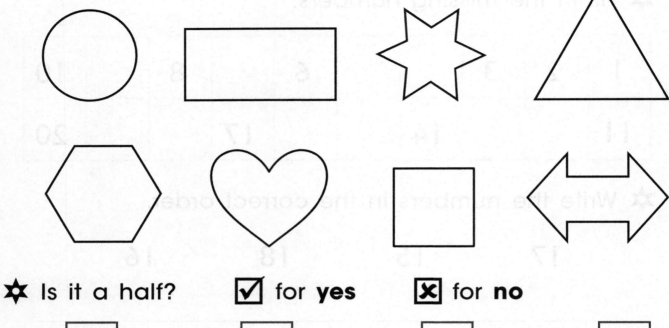

✸ Is it a half?   ☑ for **yes**   ☒ for **no**

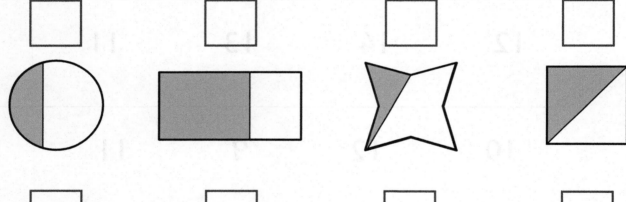

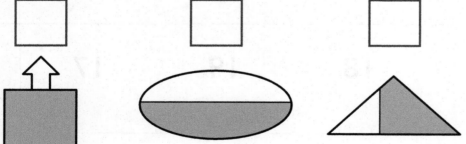

✸ Tell someone how you know that half is shaded.

Matching Halves

✿ Match the halves like this.

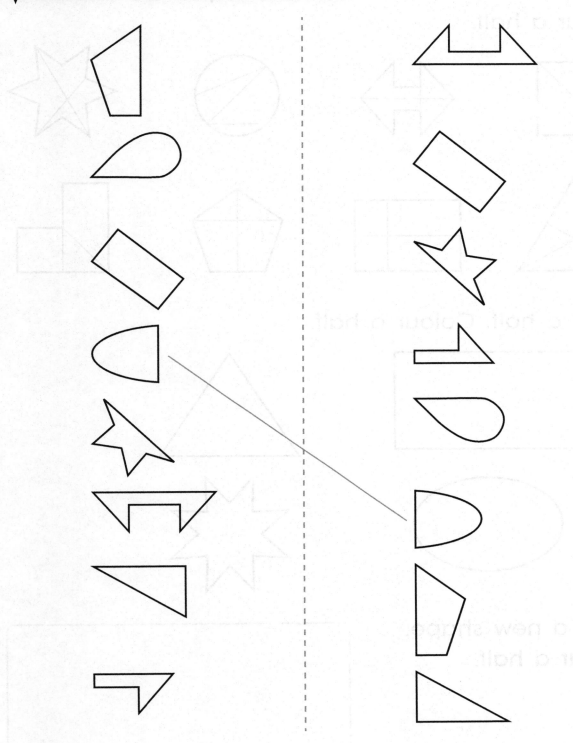

✿ Choose a colour for each matching half.

Making a Half

� Find the line that divides each shape in half.
Colour a half.

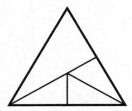 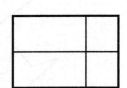  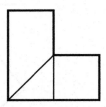

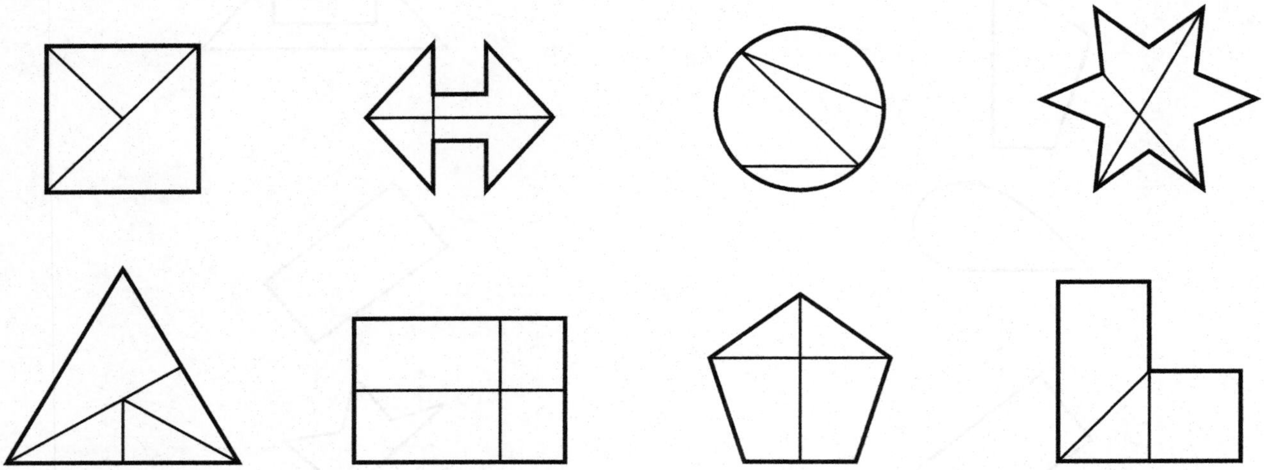

✿ Make a half. Colour a half.

 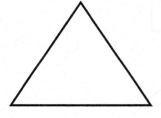

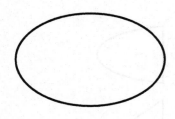 

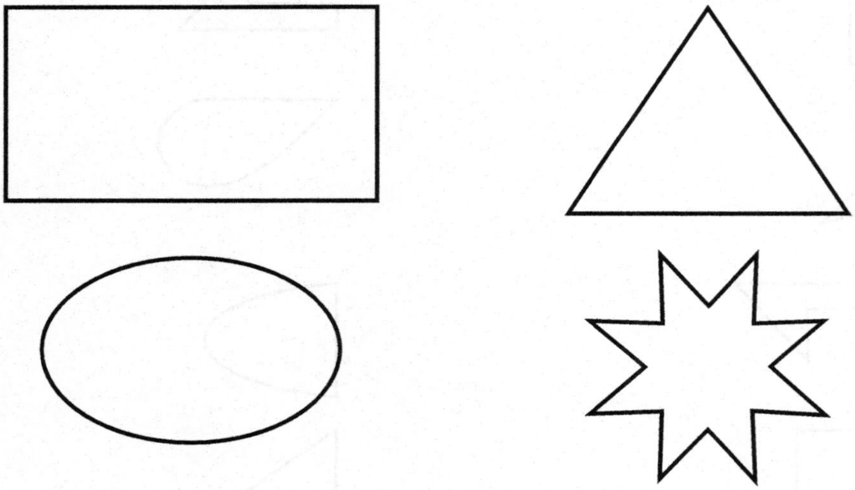

✿ Draw a new shape.
Colour a half.

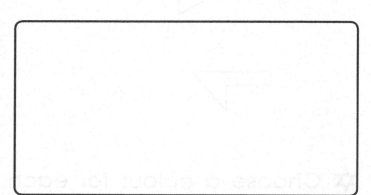

Unit **23** **Halves** (TRB pp. 110–113)
Fractions and decimals MAe-7NA describes two equal parts as halves

Unit 23

STUDENT ASSESSMENT

✹ Colour a half.

  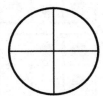 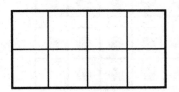

✹ Is it a half? ☑ for **yes** ☒ for **no**

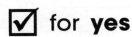

✹ Match the halves.

 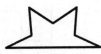 

- -

✹ How do you know that this is not
a half?

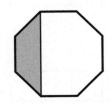

_____

Numbers on Ten Frames

✿ Write how many.

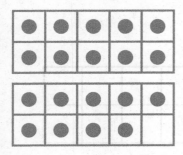

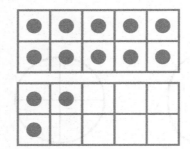

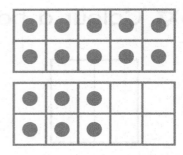

_____

_____

_____

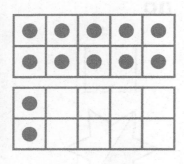

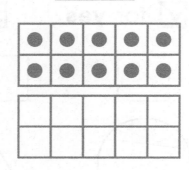

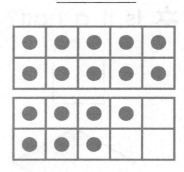

_____

_____

_____

✿ Show this many.

14

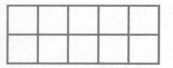

18

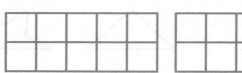

12

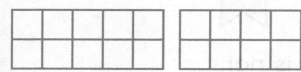

15

20

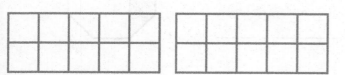

11

Numerals, Words and Pictures

Write or draw the numeral, word or picture.

11	eleven	
	twelve	
	thirteen	
	fifteen	
17		
18		
	twenty	

Unit 24 **More About Numbers to 20** (TRB pp. 114–117)
Whole numbers MAe-4NA counts to 30, and orders, reads and represents numbers in the range 0 to 20

97

Showing the Same Number

Your teacher will give you a card from
BLM 1 'Number Cards 0–5'.

Paste your number card in the middle.

Fill in the spaces around it.

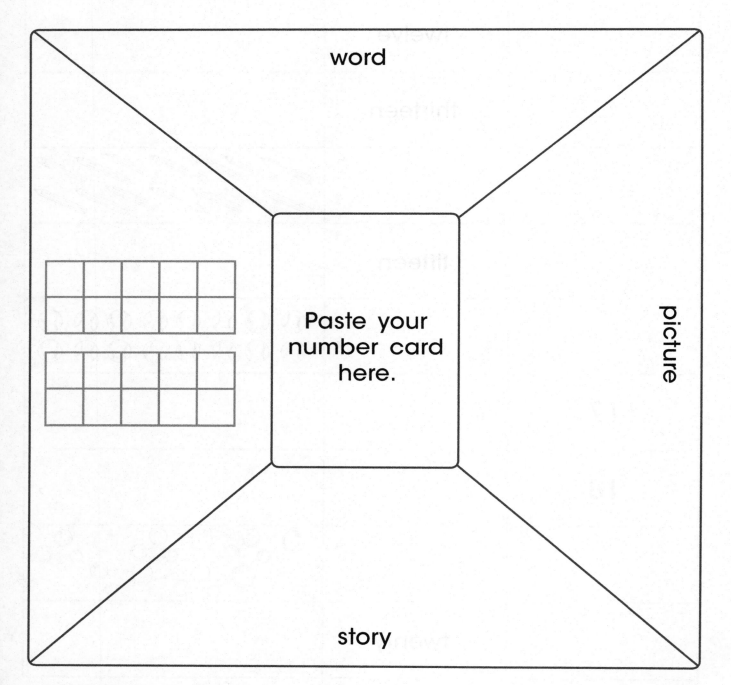

word

picture

Paste your number card here.

story

STUDENT ASSESSMENT

DATE: _____

✿ Write how many.

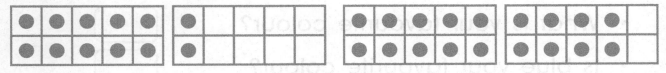

_____   _____

✿ Show the numbers on ten frames.

17   13

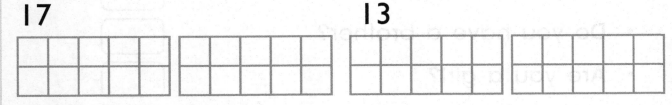

✿ Write the matching word for the number.

12 _____   17 _____

15 _____   11 _____

✿ Write the matching number.

thirteen _____   eighteen _____

sixteen _____   fourteen _____

✿ Draw lines matching pictures, words and numbers.

✔✔✔✔✔✔✔✔✔✔   ten   10

❀❀❀❀❀❀❀❀❀❀❀   sixteen   11

★★★★★★★★★★★★★★★   eleven   13

●●●●●●●●●●●●●   thirteen   16

Unit
24 **More About Numbers to 20** (TRB pp. 114–117)
Whole numbers MAe-4NA counts to 30, and orders, reads and represents numbers in the range 0 to 20

99

"Yes" and "No" Questions

✺ Read the questions. Put a ✓ if you can answer "yes" or "no".

- What is your favourite colour? ☐
- Is blue your favourite colour? ☐
- How do you come to school? ☐
- Do you walk to school? ☐
- Do you have a brother? ☐
- Are you a girl? ☐
- What do you have for lunch? ☐
- Did you have fruit for lunch today? ☐

✺ Write your own "yes" or "no" question.

_____

_____

✺ Choose one of the "yes" or "no" questions from the list. Write it here.

_____

✺ Ask two of your friends the question. Write what you find out.

_____

_____

Sandwiches for Lunch

✿ Look at the display.

Do you have a sandwich for lunch?

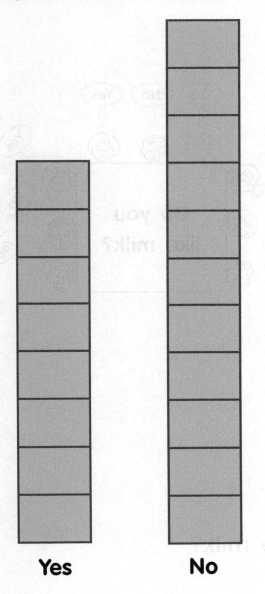

Yes   **No**

✿ Do most children have a sandwich for lunch?

_____

✿ How many children have a sandwich for lunch?

_____

✿ How many children have something else for lunch?

_____

✿ If you asked your class this question, would the display look the same?

_____

✿ Why do you think that?

_____

_____

Who Likes Milk?

✿ The teacher asked, "Do you like milk?" Draw a display to show the answers.

Do you like milk?

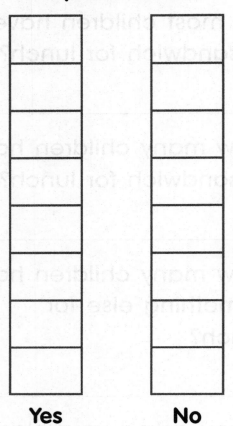

Yes   No

✿ Do most children like milk?

_____

✿ How many children don't like milk?

_____

✿ How many children like milk?

_____

✿ What if the teacher asked your class? On another sheet of paper, draw a display to show the answers.

DATE:

STUDENT ASSESSMENT

✿ ✔ the questions you can answer with "yes" or "no".

- Are your socks white?

- What colour are your socks?

- Have you been to the beach?

- Do you go to school?

- How old are you?

✿ Some children were asked "Do you have a purple pencil?"

They said:

On another sheet of paper, draw a display to show the children's answers.

Write what the display tells us.

_____

_____

Full

✦ Circle the containers that are **full**.

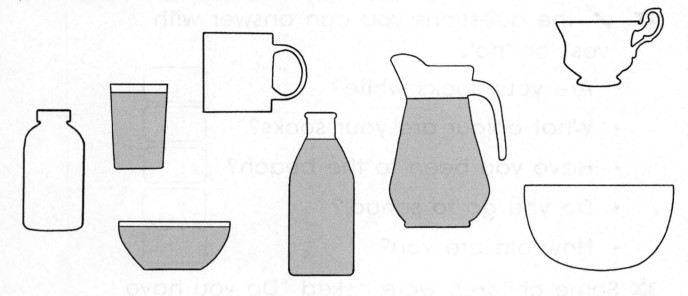

✦ Work with a partner.
Find out how many pencils
a pencil case holds.

Write what you found out.

_____

_____

_____

How Much Does It Hold? (TRB pp. 122–125)
Volume and capacity MAe-11MG describes and compares the capacities of containers and the volumes of objects or substances using everyday language

Which Container Holds More?

✸ Colour the container that holds **more**.

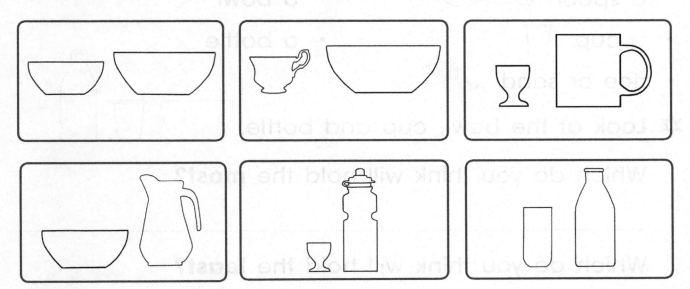

✸ How many cubes will fit into your shoe?
Have a guess first. Then check by filling
your shoe with cubes.

I guess _____ cubes.

My shoe holds _____ cubes.

✸ Find something in the
classroom that holds
more cubes than
your shoe. Draw it.

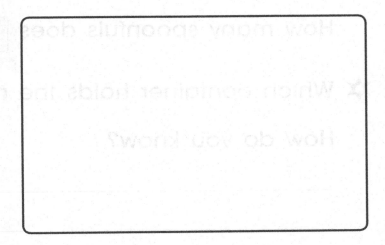

Unit
26
How Much Does It Hold? (TRB pp. 122–125)
Volume and capacity MAe-11MG describes and compares the capacities of containers and the volumes of objects or substances using everyday language

105

How Many Spoonfuls?

You will need:

- a spoon
- a cup
- rice or sand

- a bowl
- a bottle

✵ Look at the bowl, cup and bottle.

Which do you think will hold the **most**?

_____

Which do you think will hold the **least**?

_____

✵ Use the spoon and the rice (or sand) to find out how much each container holds.

How many spoonfuls does ⊔ hold? _____

How many spoonfuls does ⌣ hold? _____

How many spoonfuls does ⌷ hold? _____

✵ Which container holds the **most**? _____

How do you know?

_____

_____

How Much Does It Hold? (TRB pp. 122–125)
Volume and capacity MAe-11MG describes and compares the capacities of containers and the volumes of objects or substances using everyday language

DATE:

You will need:

- a paper or plastic cup
- counters

✪ Circle the containers that are **empty**.

  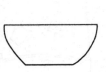

✪ Colour the container that holds **more**.

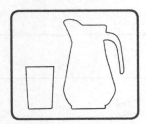 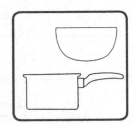  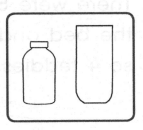

✪ Find something in the classroom that holds **more than** the paper cup.

Draw it.

How do you know that it holds **more**?

_____

✪ Guess how many counters will fit into the cup. Check by filling it with counters.

I guess _____ counters.

The cup holds _____ counters.

Unit **26** **How Much Does It Hold?** (TRB pp. 122–125)
Volume and capacity MAe-11MG describes and compares the capacities of containers and the volumes of objects or substances using everyday language

107

Teddies on the Bed

✿ **Look at the pictures and tell the story.**
One has been done.

There were **6** teddies on
the bed and **2** fell off,
so **4** teddies were left.

_____

_____

_____

_____

_____

✿ **Make your own story.**

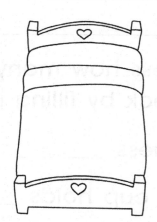

_____

_____

_____

Subtraction (TRB pp. 126–129)
Addition and subtraction MAe-5NA combines, separates and compares collections of objects, describes using everyday language, and records using informal methods

I'm Hungry

I had this many.	I ate this many.	Now I have this many.

Unit 27 **Subtraction** (TRB pp. 126–129)
Addition and subtraction MAe-5NA combines, separates and compares collections of objects, describes using everyday language, and records using informal methods

109

Teddy Bears' Picnic

✪ These teddy bears are having a picnic.

Cross out the food to show what they ate.

✪ Write about what the teddy bears ate and what was left over.

_____

_____

_____

_____

_____

Subtraction (TRB pp. 126–129)
Addition and subtraction MAe-5NA combines, separates and compares collections of objects, describes using everyday language, and records using informal methods

Unit 27

STUDENT ASSESSMENT

DATE:

✿ I had 7 nuts in a bag.

2 nuts fell out. How many do I have left? _____

✿ Complete the table.

I had this many.	I lost this many.	Now I have this many.
(9 pencils)	5	
(7 stars)	3	
(6 bears)	2	
(8 fish books)	4	
(10 apples)	8	

✿ Draw and write you own subtraction
or take-away story.

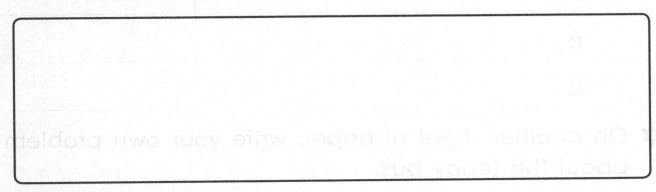

Unit 27

Subtraction (TRB pp. 126–129)
Addition and subtraction MAe-5NA combines, separates and compares collections of objects, describes using everyday language, and records using informal methods

111

Teddies on the Bus

✿ Use the bus, a dice and some counters to solve the problems.

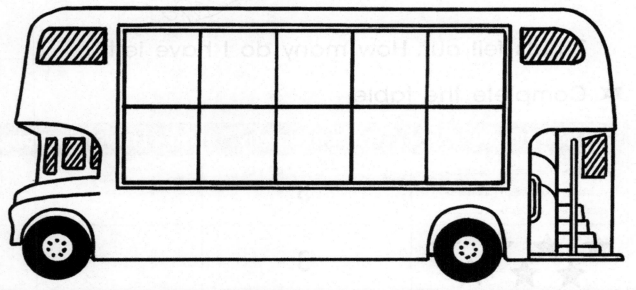

This many teddies got on the bus.	Roll the dice to find out how many teddies got off the bus	How many teddies are left?
9		
7		
8		
10		
7		
6		
8		

✿ On another sheet of paper, write your own problem about the teddy bus.

Jumping Back

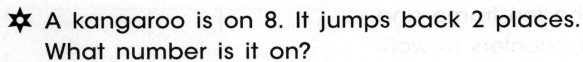

✸ A kangaroo is on 8. It jumps back 2 places.
What number is it on?

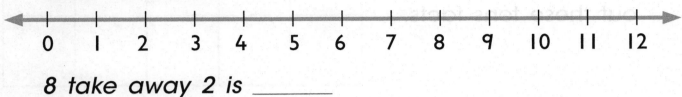

8 take away 2 is _____

✸ A kangaroo is on 10. It jumps back 6 places.
What number is it on?

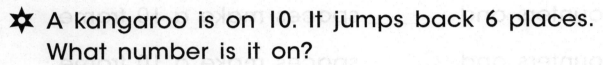

10 take away 6 is _____

✸ A kangaroo is on 7. It jumps back 3 places.
What number is it on?

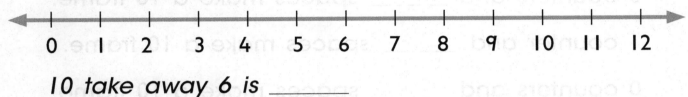

7 take away 3 is _____

✸ A kangaroo is on 9. It jumps back 4 places.
What number is it on?

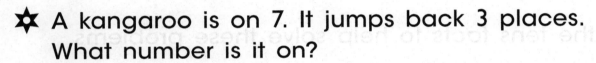

9 take away 4 is _____

✸ A kangaroo is on 6. It jumps back 3 places.
What number is it on?

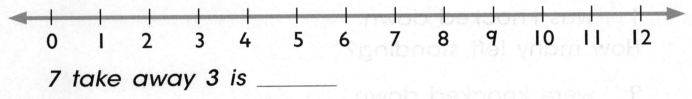

6 take away 3 is _____

Unit 28 **More About Subtraction** (TRB pp. 130–133)
Addition and subtraction MAe-5NA combines, separates and compares collections of objects, describes using everyday language, and records using informal methods

113

Tenpin Bowling

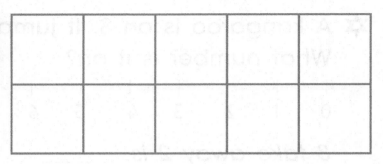

✶ Use the ten frame and some counters to work out these tens facts.

8 counters and _____ spaces make a 10 frame.

3 counters and _____ spaces make a 10 frame.

1 counter and _____ spaces make a 10 frame.

0 counters and _____ spaces make a 10 frame.

✶ Use the tens facts to help solve these problems.

1 🎳 was knocked down.
How many left standing? _____

9 🎳 were knocked down.
How many left standing? _____

3 🎳 were knocked down.
How many left standing? _____

10 🎳 were knocked down.
How many left standing? _____

More About Subtraction (TRB pp. 130–133)
Addition and subtraction MAe-5NA combines, separates and compares collections of objects, describes using everyday language, and records using informal methods

STUDENT ASSESSMENT

DATE:

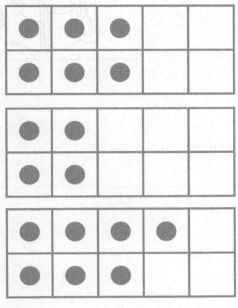

☆ I had 6 and I lost 2.
How many left? _____

I had 4 and I lost 1.
How many left? _____

I had 7 and I lost 3.
How many left? _____

☆ Look at the number line to help you solve these problems.

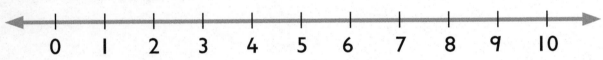

I had 8 and I gave away 3. How many left? ☐

I had 3 and I gave away 1. How many left? ☐

I had 5 and I gave away 2. How many left? ☐

I had 10 and I gave away 6. How many left? ☐

☆ Look at the ten frame. Solve these problems.

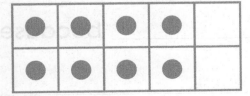

10 take away 8 is _____.

10 take away 3 is _____.

Unit
28   **More About Subtraction** (TRB pp. 130–133)
Addition and subtraction MAe-5NA combines, separates and compares collections of objects, describes using everyday language, and records using informal methods

115

My School Diary

☆ Write about and draw what you do at school during the week.

Monday
Tuesday
Wednesday
Thursday
Friday

☆ My favourite day is _____ because I

_____

What's the Time?

✷ Draw what you would be doing at:

 at night.

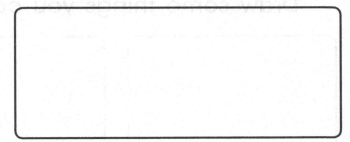

 in the morning.

✷ **What time is it?**

2:00 ☐ o'clock

 ☐ o'clock

✷ Show

3 o'clock

11 o'clock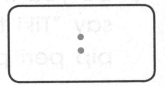

5 o'clock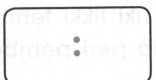

7 o'clock

In a Minute

�֎ Think about how long 1 minute is.

Draw some things you could do in 1 minute.

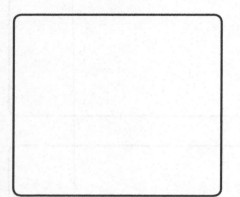 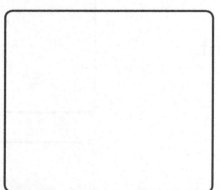 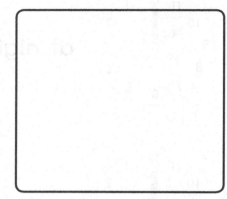

✷ Draw some things that would take you **longer** than 1 minute.

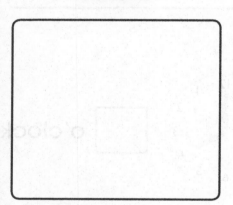 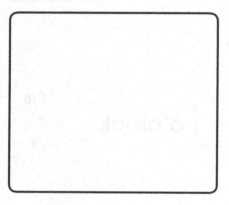 

✷ Do you think it takes **more** or **less** than 1 minute to say "Tiki tikki tembo-no sa rembo-chari bari ruchi-pip peri pembo"? _____

✷ How many times do you think you can turn around while your partner says "Tiki tikki tembo-no sa rembo-chari bari ruchi-pip peri pembo"? _____

Try it with your partner to check.

STUDENT ASSESSMENT

✺ What is the first day of the week that you come to school? _____

✺ What is your favourite day of the week?

_____

Why? _____

✺ Complete the clock.

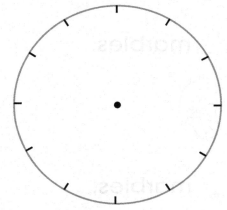

✺ Write the time for each clock.

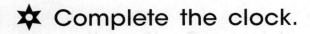

3:00

_____

✺ Show the time on the clocks.

6 o'clock

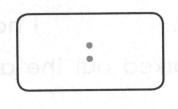

11 o'clock

4 o'clock

In My Hand

✯ Write how many marbles are in my fist.

_____ marbles.   I have 7 altogether.

_____ marbles.   I have 9 altogether.

_____ marbles.   I have 6 altogether.

_____ marbles.   I have 8 altogether.

_____ marbles.   I have 7 altogether.

✯ Write how you worked out the answers.

_____

120 **Unit 30** **Addition, Subtraction and Money** (TRB pp. 138–141)
Addition and subtraction MAe-5NA combines, separates and compares collections of objects, describes using everyday
language, and records using informal methods

Money

☆ How much do I have?

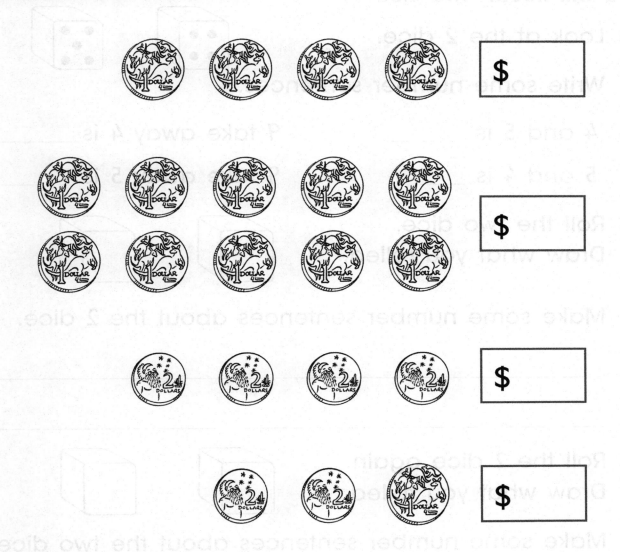

$

$

$

$

☆ Colour the coins to pay for these.

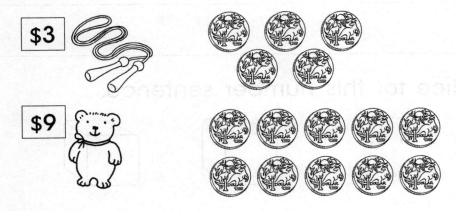

$3

$9

Unit 30

Addition, Subtraction and Money (TRB pp. 138–141)
Whole numbers (money) MAe-4NA counts to 30, and orders, reads and represents numbers in the range 0 to 20
Addition and subtraction (money) MAe-5NA combines, separates and compares collections of objects, describes using everyday language, and records using informal methods

121

Number Sentences

You will need: two dice

✪ Look at the 2 dice.

Write some number sentences.

4 and 5 is _____   9 take away 4 is _____

5 and 4 is _____   9 take away 5 is _____

✪ Roll the two dice.
Draw what you rolled.

Make some number sentences about the 2 dice.

_____

_____

✪ Roll the 2 dice again.
Draw what you rolled.

Make some number sentences about the two dice.

_____

_____

✪ Draw the 2 dice for this number sentence.

11 take away 5 is 6.

Addition, Subtraction and Money (TRB pp. 138–141)
Addition and subtraction MAe-5NA combines, separates and compares collections of objects, describes using everyday language, and records using informal methods

STUDENT ASSESSMENT

DATE:

✿ How many buttons are in the bag?

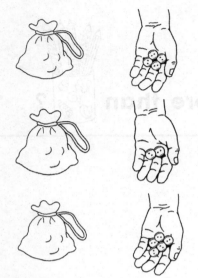

_____ buttons

I have 8 altogether.

_____ buttons

I have 6 altogether.

_____ buttons

I have 7 altogether.

✿ How much do I have?

$ ____

✿ How much do I have?

$ ____

✿ Colour the coins to pay for this.

 $5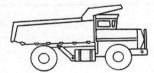

✿ What number sentence can you write using these numbers?

_____

Unit
30

Addition, Subtraction and Money (TRB pp. 138–141)
Whole numbers (money) MAe-4NA counts to 30, and orders, reads and represents numbers in the range 0 to 20

Addition and subtraction (money) MAe-5NA combines, separates and compares collections of objects, describes using everyday language, and records using informal methods

123

Maths Challenges

Challenge 1

Can you see any numbers around the room?
Copy the numbers onto paper.

Challenge 2

Can you draw a group that has **more than** [?]

How many is **less than** 1? _____

Challenge 3

Think of a number that is **more than** 5. Use a paper
plate and counters to show the different ways to make it.

Challenge 4

Think of a number
that is **more than** 10.
Can you show it in
two different ways?

Challenge 5

Make dominoes that
show 7 altogether.

Challenge 6

Draw two groups
that make 6.

Draw two groups
that make 9.

_____ and _____ is 6.

_____ and _____ is 9.

Challenge 7

Draw **two** more than 🎲 .

Draw **two** more than 1.

Challenge 8

I ate 4 cupcakes. Here are
the cupcakes that are left.

How many cupcakes did I begin with? _____

How did you work this out? _____

125

Make Your Own Game

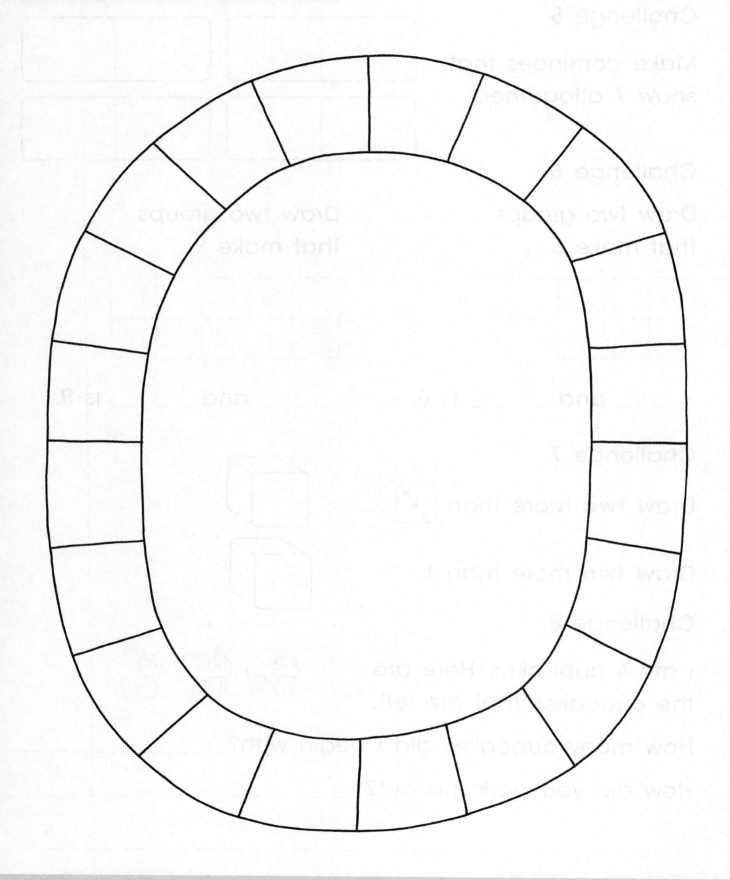

Maths Glossary

Numbers

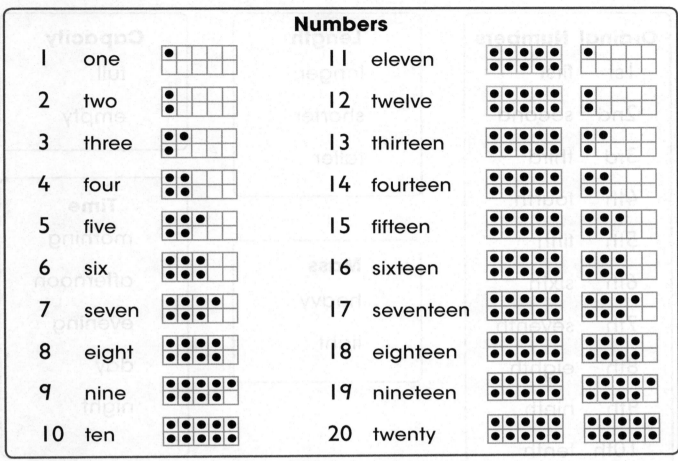

1	one		11	eleven
2	two		12	twelve
3	three		13	thirteen
4	four		14	fourteen
5	five		15	fifteen
6	six		16	sixteen
7	seven		17	seventeen
8	eight		18	eighteen
9	nine		19	nineteen
10	ten		20	twenty

2D Shapes

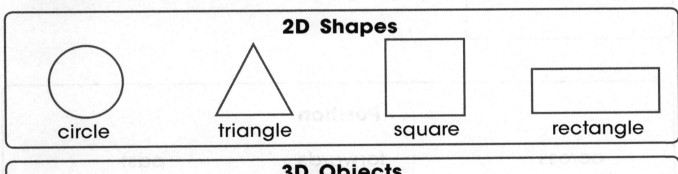

circle   triangle   square   rectangle

3D Objects

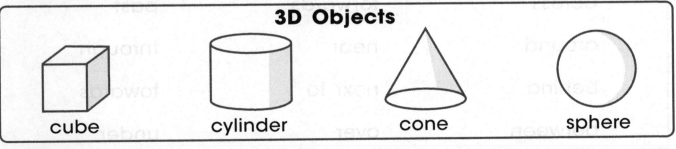

cube   cylinder   cone   sphere

Days of the Week

Sunday   Monday   Tuesday   Wednesday   Thursday   Friday   Saturday

Maths Glossary

Ordinal Numbers

1st	first
2nd	second
3rd	third
4th	fourth
5th	fifth
6th	sixth
7th	seventh
8th	eighth
9th	ninth
10th	tenth

Length

longer

shorter

taller

Mass

heavy

light

Capacity

full

empty

Time

morning

afternoon

evening

day

night

Position

across	forwards	past
around	near	through
behind	next to	towards
between	over	under
in front of		